Current Topics in Human Intelligence

Volume 5

The Environment

Current Topics in Human Intelligence

Volume 5

The Environment

Editor

Douglas K. Detterman

Case Western Reserve University

Ablex Publishing Corporation
Norwood, New Jersey

Printed in the United States of America.

ISBN:1–56750–136–2

ISSN: 8775–0040

Ablex Publishing Corporation
355 Chestnut Street
Norwood, New Jersey 07648

Contents

List of Contributors

Numbers in parentheses indicate the pages on which the author's contributions begin.

Phillip L. Ackerman, Department of Psychology, University of Minnesota, N218 Elliott Hall, 75 East East River Road, Minneapolis, MN 55455 (8)

Clancy Blair, Civitan International Research Center, The University of Alabama at Birmingham, P.O. Box 313, Birmingham, AL 35294-0021 (59, 229)

John G. Borkowski, Department of Psychology, University of Notre Dame, Haggar Hall, Notre Dame, IN 46556-5636 (3)

Chris Brand, Department of Psychology, University of Edinburgh, 7 George Square, Edinburgh EH8 9J7, United Kingdom (113)

Alan D. B. Clarke, 109 Meadway, Barnet, Herts EN5 5JZ, United Kingdom (121)

Ann M. Clarke, 109 Meadway, Barnet, Herts EN5 5JZ, United Kingdom (121)

Tammy L. Dukewich, Department of Psychology, University of Notre Dame, Haggar Hall, Notre Dame, IN 46556-5636 (3)

James R. Flynn, Department of Political Studies, University of Otago, P.O. Box 56, Dunedin, New Zealand (2, 197)

Howard L. Garber, Milwaukee Center for Independence, 1339 North Milwaukee Street, Milwaukee, WI 53202 (127)

James Hodge, 431 West Maine St., Madison, WI 53703 (127)

Arthur R. Jensen, School of Education, University of California, Berkeley, CA 94720 (147)

Daniel P. Keating, Centre for Applied Cognitive Science, The Ontario Institute for Studies in Education, 252 Bloor Street West, Toronto, Ontario M5S 1V6 Canada (31, 211)

John C. Loehlin, Department of Psychology, University of Texas, Austin, TX 78712 (151)

Richard Lynn, University of Ulster at Coleraine, Coleraine, County Londonderry, BT52-1SA Northern Ireland (157)

Thomas G. O'Connor, Curry Programs in Clinical and School Psychology, University of Virginia, 147 Ruffner Hall, 405 Emmet Street, Charlottesville, VA 22903-2495 (45, 217)

William L. Oliver, Department of Psychology, Florida State University, Tallahassee, FL 32306-1051 (87, 247)

Robert C. Pianta, Curry Programs in Clinical and School Psychology, University of Virginia, 147 Ruffner Hall, 405 Emmet Street, Charlottesville, VA 22903-2495 (45, 217)

Ernesto Pollitt, Behavioral Research and International Nutrition, Department of Pediatrics, Univeristy of California, Davis, CA 95616-8538 (163)

Craig T. Ramey, Civitan Internation Research Center, The University of Alabama at Birmingham, P.O. Box 313, Birmingham, AL 35294-0021 (59, 229)

Carmen Saco-Pollitt, Department of Teacher Education, California State University, Sacramento, CA 95819-6079 (163)

Herman H. Spitz, 389 Terhune Road, Princeton, NJ 08540 (173)

Lee Anne Thompson, Department of Psychology, Case Western Reserve University, 10900 Euclid Avenue, Cleveland, OH 44106-7123 (179)

Theodore D. Wachs, Department fo Psychological Sciences, Purdue University, 1364 Psychological Sciences Building, West Lafayette, IN 47907-1364 (69, 235)

Richard K. Wagner, Department of Psychology, Florida State University, Tallahassee, FL 32306-1051 (87, 247)

Douglas Wahlsten, Department of Psychology, University of Alberta, Edmonton, Alberta T6G 2E9, Canada (185)

Richard A. Weinberg, Institute of Child Development, University of Minnesota, 51 East River Road, Minneapolis, MN 55455-0345 (191)

Current Topics in Human Intelligence Foreword to the Series

Douglas K. Detterman, Editor

The purpose of this series is to focus on single issues of importance to the study of human intelligence. Unlike many edited volumes, this one is designed to be thematic. Each volume will present a detailed examination of some question relevant to human intelligence and, more generally, individual differences.

The reason for beginning this series is that a forum is needed for extensive discussion of pertinent questions. No such forum currently exists. A journal does not allow an author sufficient space or latitude to present fully elaborated ideas. Currently existing edited series are not thematic but, instead, offer researchers an opportunity to present an integrative summary of their own work. Both journals and edited, nonthematic monographs are essential to the advancement of the study of human intelligence. But they do not allow collective intelligence to be brought to bear on a single issue of importance.

Why is it important to examine specific issues in detail? The answer to this question depends on an appreciation of the historical development of the study of human intelligence. For at least 40 years, and perhaps longer, the study of human intelligence has been less than highly regarded as an academic pursuit. The reasons for this attitude are many and have been discussed elsewhere. Despite the reasons, the lack of academic sanction for the study of individual differences in human intelligence has produced a discipline without a unifying paradigm. When researchers study human intelligence, it is almost always from the perspective of training in a related, but different, discipline. Disciplines that have "sacrificed" researchers to human intelligence include: cognitive, developmental, and educational psychology, behavior genetics, psychometrics, mental retardation, neuropsychology, and even experimental psychology. Each of these researchers has brought a different set of assumptions and methods to the study of human intelligence.

That so many different points of view have been applied to a single subject area has, in my opinion, brought vitality to the endeavor, a vitality currently lacking in so many areas of the social sciences. But this vitality does not arise from the isolation and fractionation that can be the result of different points of view. On the contrary, it results from the juxtaposition of these different points of view, a juxtapositioning that has occurred with increasing frequency over the last 10 years.

Therefore, it is the purpose of this series to bring different points of view together on issues of importance to understanding human intelligence. The hope is that, at the very least, researchers will find more reason to give their primary allegiance to the study of human intelligence and, at the most, the series will contribute to the emergence of a unifying paradigm.

Foreword To Volume 5

This is the fifth volume in the series. Each volume has been dedicated to a specific topic concerning human intelligence. Specialists in a topic were asked to write chapters and the chapters were published. That is the usual format of an edited book. This volume is different.

Volume 5 is dedicated to the environment's effect on intelligence, one of the most contentious issues in intelligence research. Indeed, the nature–nurture issue has been one of the most debated in psychology. It seemed inappropriate to have researchers just write chapters about such a hotly contested issue.

Therefore, the book is divided into three sections. As a beginning point, a few people were asked to write chapters on the environment's contribution to intelligence. A larger number of persons were then given the original chapters and asked to write shorter papers. Authors of the shorter chapters (in Part 2 of the book) were told to write about any topic they thought important. Their chapter could be a response to a single chapter, a response to a point made in several chapters, or a brief statement on their position about the environment and intelligence. Each option was used by at least one author. A third section then contains rebuttals written by authors of the original seven chapters. Giving these authors the opportunity to address the concerns of the commentaries allows for a well-rounded forum of discussion.

In reading the three parts of this volume, I think the reader will find the format a particularly stimulating one for this topic. It captures the intensity of debate about the environment's effects on intelligence. The format conveys the lack of agreement on findings and methodology and highlights the important issues where there is at least some agreement among some factions.

What this format fails to do is give the reader everything they want. I would have liked two or three sets of exchanges between these sets of authors. Unfortunately, this was impossible for several reasons. First, the book would have ended up as several books. Second, because each exchange takes 6 to 9 months to complete, by the time several exchanges had been completed, the earlier work would be 3 or 4 years out of date.

Has this book done anything to resolve the debates about intelligence and the environment? Because the effect of the environment on intelligence is one of the oldest debates in the history of psychology and education, it would be presemptuous to think that a single book, even a book with a unique format, could resolve any of the long-standing issues in this field. Despite the long history of unresolved conflict, I do think that this book brings into focus some important issues for understanding the relationship between environment and intelligence. The reason for this isn't the book or the format—it is simply timing. Empirical data important to this controversy have recently emerged to cast new light on the issues. Those data are fully reflected in this volume.

The other obvious point that emerges from this book is the complexity of the relationship between environment and intelligence. Environmental variables are many and, individually, their effects seem small. The question of how much environment influences intelligence will be easier to answer than how environment influences intelligence. The hardest question of all to answer will be how our knowledge of environment can be used to systematically improve education and, perhaps, increase intelligence.

Whatever else this book may or may not be, it is a demonstration that the nature–nurture issue still has the potential to excite researchers who are interested in human intelligence.

Douglas K. Detterman
May 5, 1994

Part 1

Current Topics on Human Intelligence: The Environment

Chapter 1

Environmental Covariations and Intelligence: How Attachment Influences Self-Regulation*

John G. Borkowski
Tammy L. Dukewich

University of Notre Dame

It is difficult to define, isolate, and measure unique components of the environment with conceptual and psychometric precision. One problem is that environments are in a constant state of flux. Furthermore, individuals' perceptions of the static or dynamic qualities of environments are often not identical, even when environmental consistency is high such as for members of a stable family unit. Hence, it is not easy to develop precise psychological theories about how specific aspects of an environment, or aggregated environmental components, influence the development of cognition and intelligence. Despite

* The writing of this chapter was supported by NIH Grant HD–25456. We are appreciative of the assistance of Ellen Moss and her colleagues in providing prepublication materials.

these and other conceptual and methodological problems, recent paradigmatic advances have given developmental psychologists a more solid foundation on which to address questions about how environments are created and modified, and how their salient features covary to influence a wide range of human behaviors.

Wachs' (1992) recent analysis of the nature and function of environments represents a meaningful advance in paradigmatic perspective. His call is for examining variability in development that results from the interaction of different levels, or kinds, of environmental dimensions together with specific organismic–genetic factors, all carried out within a longitudinally-based process orientation. More specifically, Wachs (1992) claimed that psychosocial environments covary with salient child characteristics and act in combination to influence human development differentially across time. We will use this framework to propose a model that describes the emergence of self-regulation and its long-term causal links to intelligence, as viewed from information processing and psychometric perspectives.

We focus on environmental covariations that reside in the attachment history of each child and exert their influence on intellectual development through the emergence of self-regulatory skills. We hypothesize that the nature and quality of instructional environments that promote regulation vary with attachment status which, in its own right, directly influences attentional flexibility, resulting in additional variability in emotional and cognitive regulation. Eventually, the development of self-regulatory skills influences the formation of selected aspects of the knowledge and skill represented in childhood intelligence.

A MODEL OF ATTACHMENT-RELATED CAUSES OF REGULATION

Figure 1 presents a general model of the causes of attachment, as well as its cognitive–intellectual consequences. On the left side of the figure are some of the distal and proximal causes of infant attachment. Most prominent is the role played by the sensitivity and responsivity of the child's caregiver who is often, but not exclusively, the mother. These qualities of the infant's immediate environment are assumed to be pivotal in the development of attachment security, and are linked to maternal characteristics such as cognitive readiness for parenting, reconciliation of prior attachment history, psychological stability (e.g., depression), and personal resources (e.g., "hardiness") (Colin, 1991). Additionally, these environmental factors are associated with more general contextual factors such as the availability of instrumental and emotional supports from family and peers and the quality of neighborhoods (Borkowski, Whitman, Passino, Rellinger, Sommer, Keogh, & Weed, 1992; Brooks-Gunn, &

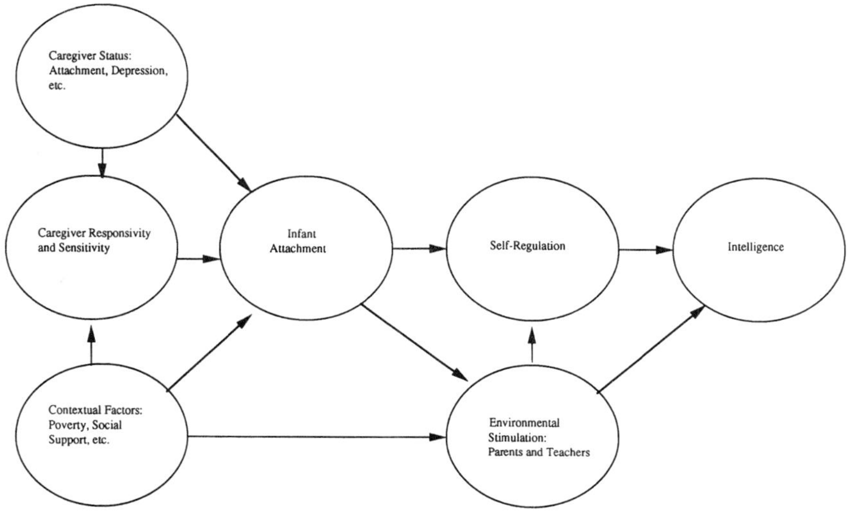

Figure 1. A model of the causes and consequences of attachment and its direct and indirect influences on self-regulation and IQ.

Kato, 1991; Nath, Borkowski, Whitman, & Schellenbach, 1991). These macro- and molar-level variables are known to have direct and/or indirect effects on infant attachment (Colin, 1991). It should be emphasized that there is little solid evidence pointing to a definitive relationship between attachment security and the emergence of intelligence (Lamb, Thompson, Gardner, & Charnov, 1985). It is the purpose of this paper to suggest missing links in the chain of events that flow from infant attachment to childhood intelligence.

Inspired by the insightful ideas of Main (1991), we have begun to consider whether, and how, insecure attachment might delay the emergence of self-regulation and eventually produce deficits in intelligence. We do not consider attachment as a sufficient construct that, in isolation, accounts for advanced forms of cognition and complex components of intelligence, but rather that it may be one of several salient organismic variables (temperament being another likely candidate) that directly influences the formation of self-regulation and, more interestingly, determines the quality of environments in which control and monitoring skills are modeled and taught.

Self-Regulation and Intelligence

Although the claim that self-regulation is at the heart of human intelligence is not universally accepted, it is assumed by many to be critical for most problem-solving and reasoning tasks and, presumably, plays an instrumental role in knowledge accumulation and domain-specific skill acquisition. Sternberg (1985), Mithaug (1993), Butterfield and Albertson (1993), Reid and Borkowski (1987) and others have argued that executive processing, higher-order metacognition, and self-control (terms often used synonymously for self-regulation) are essential for complex learning and cognition. Historically, self-regulation is seen as the hallmark of intelligence (Brown, 1987). The basic argument is that regulatory processes permit task analysis, choices about tactical and strategic approaches, active processing of performance feedback, and, finally, strategic revision, problem reformulation, and/or the recognition of a correct solution. In this way, what we know and how we learn are associated with the quality of our self regulatory skills.

Attachment and Regulation

As stated previously, we hypothesize that an infant's attachment status has both direct and indirect effects on the later development of self-regulation. The direct organismic effect is assumed to be mediated by attentional flexibility and early emotional regulation: The securely attached child who explores the environment, and who controls emotions in the face of stress and challenge, has the basic prerequisites for learning how to perform task analyses, how to be planful, and how to respond with deliberation to obstacles and frustration encountered during complex learning opportunities.

Indirectly, securely attached children help create enriched environments that foster the growth of self-regulation. These stimulating environments encourage exploration, innovation, and the emergence of creativity. Prompted, in large part, by the skills and dispositions of the securely attached child, an enriched environment will likely be characterized by more stimulating learning activities and by more challenging classroom interactions and expectations, all of which facilitate self-regulatory and intellectual development. In the next section, we present preliminary evidence on the plausibility of the attachment-regulation hypothesis and its consequences for intellectual development.

COGNITIVE–INTELLECTUAL CONSEQUENCES OF ATTACHMENT

Direct Effects on Regulation: "Missed" Learning Opportunities

Although the causal influence of early attachment on regulation (and other aspects of cognition, for that matter) has not been clearly demonstrated, we hypothesize that attachment security is intimately related to a preference for novelty, sustained attention, and the early appearance of emotional regulation that, in concert, lay the seeds for metacognitive understanding and the groundwork for control and monitoring skills (cf. Butterfield & Albertson, 1993). In contrast, attentional resources are often occupied as insecure infants attempt to cope with their attachment relationships in stressful situations, leaving fewer resources and perhaps less capacity for exploring and learning about new aspects of the environment (cf. Main, 1991).

In a series of experimental and correlational studies, Dunham and Dunham (1990) found that interactions between infants and their caregivers begin to influence the quality of infants' attention to relevant aspects of the environment as early as 3 months of age. More specifically, the more time a mother–infant pair spends in vocal turn-taking, the longer infants will fixate on a subsequent stimulus pattern and the shorter their interfixation intervals during a contingency task, resulting in increases in social interactions. It may be that interchanges that facilitate the growth of attention also promote the formation of attachment security.

Borkowski et al. (1992) have recently examined concurrent relationships between the security of infants' attachment to their mothers and the flexibility of their toy play, and subsequently linked these variables to intellectual development at age three. We expected that differences between secure and insecure infants would be particularly pronounced during the reunion episodes of the Strange Situation procedure (Ainsworth & Wittig, 1969). For secure infants, the caregiver's return was anticipated and expected; once reunited and comforted, secure infants should be free to return their attention to their environment. In contrast, insecurely attached infants, who generally experienced greater stress and were not easily comforted following separation episodes, were expected to experience cognitive and/or emotional confusion upon a caregiver's return, resulting in less flexible attention to novel stimuli and fewer meaningful interactions with their immediate environment.

These hypotheses were tested with a sample of 59 first-time adolescent mothers and their infants. Infants were videotaped in the Strange Situation paradigm when they were approximately 12 months of age. Forty-two percent were categorized as secure, while the other 58 percent were categorized as insecure, generally anxious–avoidant or disorganized. In addition to scoring the videotapes according to standard attachment procedures, each tape was also coded to determine the flexibility of attention to the toys present in the room. The point biserial correlation between security of attachment and attentional flexibility (during the first reunion episode) was significant ($r = .27$). More interestingly, attentional flexibility correlated with Bayley mental development at one year ($r = .45$), Stanford-Binet IQ at three years ($r = .47$), and receptive language (PPVT-R) at three years ($r = .48$). It appears that the ability of infants to attend selectively to relevant aspects of their environments, and to adapt flexibly to changes in environments that are associated with stress and challenge, are critical components of cognitive–intellectual development (cf. Columbo & Mitchell, 1990).

The relationship between attentional deployment and measures of intelligence were also examined by Rodriguez, Mischel, and Shoda (1989), using a delay of gratification task. Children who were able to redirect their attention away from an object of gratification were better able to delay gratifications and scored higher on verbal intelligence (PPVT). Delay of gratification and attentional deployment may be part of a more comprehensive construct of self-regulation, which could be the actual correlate or cause of intelligence: "Effective attention deployment is a basic ingredient of the ability to exercise goal-directed self-control" (Rodriguez et al., 1989, p. 365).

These studies provide preliminary support for our hypothesis about the direct effects of attachment on the emergence of cognitive control, which in turn contributes to the development of intelligence. In stressful situations, infants, and perhaps children up to 3–4 years of age with anxious attachments (Crittenden, 1992), may be compelled to direct their attentional resources to unresolved emotional dilemmas. As a result, fewer resources are available for exploring, and learning from, important aspects of their environments, thus impeding the development of attentional processing. Although early noncontingent responsivity between mothers and infants may provide a partial explanation for both insecure attachment as well as failures in the regulation of attention, we believe the emergence of insecure attachment further interferes with the development of attentional flexibility, thus inhibiting the formation of critical self-regulatory skills.

Indirect Effects of Attachment on Self-Regulation: Creating Stimulating Environments

Although we know little about why and how stimulating environments are created, sustained, and modified, genetically-based organismic factors apparently account for a considerable portion of the variance in children's environments (Plomin & Bergeman, 1991). In this section we present data suggesting that other early appearing and relatively stable child characteristics (such as attachment) also influence the quality and functioning of critical aspects of environments which continue to modulate the growth of self-regulation. More specifically, we argue that infant attachment exerts its indirect influence on the development of self-regulation by shaping the instructional goals, methods, and expectations of parents and teachers (see Figure 1).

Attachment and the focus of parental instruction. Based on the important data of Ellen Moss and her colleagues at Montreal, it appears that parents with similar IQs and educational levels instruct young children differently depending upon their children's attachment status. This work is based on the premise that strategic problem solving activities enter a child's zone of proximal development between the ages of three and five and that language serves as a scaffolding mechanism in parent–child collaborative interactions (Moss, Gosselin, Parent, & Dubeau, 1993; Parent, Moss, Gosselin, & St. Laurent, 1993).

In the context of meaningful interactions, young children need and accept their mothers' guidance in acquiring early metacognitive skills and understanding. Mothers of securely attached children are especially sensitive to ceding control to their children as they begin to spontaneously utilize skills and strategies previously modeled by the mother (Moss et al., 1993). We believe that during this period a spurt in cognitive self-regulation occurs as each child internalizes metacognitive understanding, thus providing opportunities for the acquisition of higher-order planning, monitoring, and analytic skills.

Moss et al. (1993) hypothesized that mothers of securely attached children are more likely to model regulatory skills and to demonstrate or prompt more sophisticated use of higher-level strategies during collaborative interchanges. Not only was this hypothesis supported by the data, but maternal interventions with secure children were also found to be finely tuned to children's previous learning histories: Following an error, instructions became more concrete, and following success they became supportive of higher-level strategies. The long-term outcomes of such flexible, adaptive instruction are likely to be greater independence, autonomy, and creativity in formal and informal learning opportunities. In contrast, exchanges between insecurely attached children and their mothers were more likely to center on lower-level activities and avoid the prompting of metacognitive understanding and monitoring. Moss et al. (1993) maintained that these differences in shared metacognitive activities during the preschool period—attributable in part to attachment status—may differentially prepare children for meeting the intellectual challenges of formal schooling.

Attachment and classroom experiences. Although Moely, Santulli, and Obach (1993) have reported that teachers generally do little to promote cognitive self-regulation through explicit instructions, a number of recent instructional programs (such as Brown's community of learners and Bransford's video technology approach) appear to focus intentionally on the teaching of problem-solving skills, including self-regulation. We know little, however, about how the insecurely attached child will fare in these enriched learning environments.

We suspect, based on the research of Sroufe and Egeland (1991), that attachment status will influence how teachers and peers interact with, and respond to, insecurely attached children: Those with long histories of insecure attachment have been shown to be more socially isolated, even rejected, by peers in the same classroom. Interestingly, preschool teachers tended to treat secure and insecure children differentially. Teachers were warmer and more age-appropriate with secure children, encouraging self-directed, autonomous learning. With anxious–resistant children, teachers were warm but controlling, thus restricting their metacognitive learning opportunities. A similar pattern of interchange was noted by Moss and colleagues (1993) in observing how mothers teach insecurely attached children. Finally, Sroufe and Egeland (1991) suggested that teachers of avoidant children were generally neither warm nor supportive, even showing emotional outbursts during the course of their interactions. These patterns of instructional styles and goal orientations seem to place the securely attached child at a distinct advantage in terms of acquiring beliefs and dispositions necessary for actualizing self-regulation.

If similar adult–child interchanges persist in the home and classroom throughout the early elementary school years, we would expect that extremely low levels of metacognition will be associated with early appearing and persistent insecure attachment. In this way, insecure attachment may have subtle but pervasive effects on cognitive development by not only creating, but also by maintaining, less stimulating and challenging learning environments. These long-term negative effects of insecure attachment on cognition and intelligence would be exacerbated if a child is surrounded by poverty, marital instability, or mental illness in the immediate family, contextual factors operating at a macro level that influence the key direct and indirect mechanisms in the proposed model (see Figure 1).

CAVEATS AND CONCLUSIONS

We have attempted to advance an interactive hypothesis that accounts for variations in attachment, self-regulation, problem solving, and intelligence in children who are presumably not characterized by strong genetic differences. Although this latter assumption is difficult to substantiate, it does not appear that patterns of attachment have strong heritability components except insofar as temperament and attachment are causally related. Instead, attachment theorists generally maintain that intergenerational transmission occurs through the passing on of supportive or maladaptive parenting practices (Colin, 1991), and we have reflected this view in our model.

Attachment, Regulation, and Intelligence: Interactive Mechanisms

The model we propose assumes that self-regulation is a complex process involving the interrelationship of lower-level strategy skills, higher-level metacognitive understanding and control, and motivationally based goals and beliefs (Reid & Borkowski, 1987). These constructs contain specific learning strategies, mechanisms that allow for strategy choices, judgments about current learning states, a willingness to define future goals, and an ability to analyze to-be-learned tasks and monitor performance. Furthermore, it is the integration of these lower- and higher-level skills, together with the motivational components associated with goal orientation and achievement, that represent our view of self-regulation. In this sense, it is somewhat similar to Mithaug's (1993) recent perspective on self-regulation and intelligence: "Persons who consistently maximize their gain across goal and choice contingencies appear to be more intelligent than those who only occasionally succeed, and . . . they feel more intelligent, too" (p. 159). Intelligent behavior is thus represented by effective self-regulation, accompanied by a clear recognition and appreciation of its significance.

Given the complexity of cognitive self-regulation, it is not surprising that multiple components influence its development. We have proposed only one causal factor—attachment—in our analyses of the self-regulation–intelligence linkage. Other constructs or processes—such as the formation of attributional beliefs, achievement goals, acquisition of specific learning skills, and processing speed—represent additional influences. Within the attachment process itself, one strand of events seems related to attentional processing, providing the basic foundation necessary for acquiring task analytic skills and monitoring abilities, especially in the face of complexities and distractions. This attentional component is likely to be more severely disrupted in Type C (ambivalent) and Type D (disorganized) children, some of whom will be later classified as coercive in early childhood (Crittenden, 1992).

The other side of attachment—its indirect instructional effects—serves to promote or restrict learning experiences, out of which grow metacognitive understanding and the ability to monitor and control cognitive actions (Butterfield & Albertson, 1993). A key process, related to the direct effect discussed above, is acquiring an awareness of when to be reflective as well as how to allocate available attentional and strategic resources. Relatedly, the adoption of task-oriented (vs. ego-oriented) learning goals, as well as beliefs about self-efficacy, presumably underlie the smooth functioning of self-regulated behaviors (Nicholls, 1989). We believe that these motivational states energize self-regulation and arise out of rich, supportive instructional environments that probably occur repeatedly in new and interesting ways over long periods of time.

The net result of acquiring self-regulating skills—in part through the proposed attachment-driven chain of events—is the potential for independent and creative learning. A planful child, blessed with self-regulatory skills, is in a position to learn more efficiently and profoundly in many academic domains. As a result, the knowledge and skill measured on psychometrically-defined IQ tests should be significantly elevated for those children who have shown a consistent pattern of self-regulated learning.

On Effect Sizes and Modifiability

Although we have presented preliminary research findings which support the plausibility of the proposed relationships among attachment, self-regulation, and intelligence, we are aware of the limited data on the impact of attachment on intellectual development. Our best guess is an effect size in the magnitude of 5 to 10 percent. A study by van Ijzendoorn and van Vliet-Visser (1988) supported the claim of a modest relationship between IQ and attachment. Interestingly, these investigators found that securely attached children displayed greater enthusiasm about problem solving, more active exploration of their environments, greater ego control, and a proficiency for overcoming frustration. These qualities are the presumed characteristics of self-regulation, as mentioned earlier, and indirectly endorse its importance in explaining the more general attachment–intelligence relationship.

It is likely that variations in intelligence will be more highly associated with self-regulation than with attachment status. Given the more direct and immediate connections between regulation and intelligence, we speculate that 15 to 20% of the IQ variance in children aged 10–14 might be due to differences in self-regulatory skills. This suggests that factors in addition to attachment influence the development of regulation and that other variables, such as speed of processing and working memory, play major roles in the determination of intelligence.

There are three points about the triadic relationship involving attachment, regulation, and IQ that are linked to the issue of modifiability. First, attachment security is not necessarily stable: A child on the path to becoming insecurely attached may become securely attached if a caregiver becomes more responsive and sensitive (Sroufe & Egeland, 1991). This shift in attachment status would allow the child to use attentional resources more actively (direct effect) and appropriately and would also help to restructure his or her immediate environment (indirect effect). Second, cognitive (and affective) self-regulation can be modified. That is, children taught to monitor and evaluate their own performance during problem solving generally increase their self-regulation skills. For instance, parents and teachers of insecurely attached children might use scaffolding methods, typically employed by the primary caregivers of securely attached children, to construct situations where children practice self-regulation and acquire metacognitive strategies following problem-solving successes (Moss et al., 1993). It is possible that this approach may be differentially effective with insecurely attached children—with Type A children benefitting more than Types C and D. Third, securely attached children who have rudimentary self-regulatory skills may, nevertheless, show non-optimal intellectual development if their environments are deficient in providing learning opportunities in which regulatory skills become practiced, refined, and expanded. It is important to remember that a necessary cause of early intelligence, such as attachment status and its regulatory counterparts, is probably not a sufficient cause that will dictate the eventual level of intellectual development. As suggested, only a portion of the variance in regulation is accounted for by attachment, and, similarily, only a portion of the variance in intelligence is accounted for by self-regulation.

SUMMARY

The model we have proposed reflects diverse contributions to the formation of attachment patterns. More interestingly, it postulates direct and indirect effects that flow from attachment to self-regulation and thereby influence cognitive–intellectual development. The most important of these factors are the proximal impact on regulation of attentional resources and the remote influence on instructional goals, methods, and styles. These direct and indirect effects can presumably be attenuated or exacerbated by other environmental events, such as the stability and quality of familial and neighborhood networks.

In several ways, our model mirrors Wachs' (1992) prescription for Phase III environmental research: Conduct theoretically-guided studies of organismic and environmental covariations, that change across time and that are influenced by other psychological, biological, and social events operating at both micro and macro levels. Meeting this research challenge effectively will certainly require the full range of theoretical and methodological ingenuity from psychologists interested in the formation and modification of intelligence from an environmental perspective.

REFERENCES

Ainsworth, M. D. S., & Wittig, B. A. (1969). Attachment and exploratory behavior of one-year-olds in a strange situation. In B. M. Foss (Ed.), *Determinants of infant behavior* (pp. 113–136). London: Metheun.

Borkowski, J., Whitman, T., Passino, A., Rellinger, E., Sommer, K., Keogh, D., & Weed, K. (1992). Unraveling the "New Morbidity": Adolescent parenting and developmental delays. In N. W. Bray (Ed.), *International review of research in mental retardation* (Vol. 18, pp. 159–196). New York: Academic Press.

Brooks-Gunn, J., & Kato, P. (1991). The impact of neighborhood poverty on three-year-old black children's IQ and home environment. In *New directions in child and family research: Shaping head start in the 90s* (pp. 187–188). Washington, DC: Department of Health and Human Services.

Brown, A. L. (1987). Metacognition, executive control, self-regulation and other more mysterious mechanisms. In F. E. Weinert & R. H. Kluwe (Eds.), *Metacognition, motivation and understanding* (pp. 65–116). Hillsdale, NJ: Lawrence Erlbaum.

Butterfield, E. C., & Albertson, L. R. (1993, June). *Metacognitive understanding, monitoring and control as determinants of cognition.* Paper presented at a conference on Research on Memory Development: State-of-the Art and Future Directions. Castle Ringberg, Germany.

Colin, V. (1991). *Human attachment: What we know now* (Contract No. HHS–100–90–00). Washington, DC: Office of Planning and Evaluation, Department of Health and Human Services.

Columbo, J., & Mitchell, D. N. (1990). Individual differences in early visual attention: Fixation time and cognitive processes. In J. Columbo & J. Fagan (Eds.), *Individual differences in infancy: Reliability, stability, prediction* (pp. 193–228). Hillsdale, NJ: Lawrence Erlbaum.

Crittenden, P. M. (1992). Quality of attachment in the preschool years. *Development and Psychopathology, 4*, 209–241.

Dunham, P., & Dunham, F. (1990). Effects of mother–infant social interactions on infants' subsequent contingency task performance. *Child Development, 61*, 785–793.

Lamb, M. E., Thompson, R. A., Gardner, W., & Charnov, E. L. (1985). *Infant–mother attachment: The origins and developmental significance of individual differences in Strange Situation behavior*. Hillsdale, NJ: Lawrence Erlbaum.

Main, M. (1991). Metacognitive knowledge, metcognitive monitoring and singular (coherent) vs. multiple (incoherent) models of attachment: Findings and directions for future research. In C. Parkes, J. Stevenson-Hinde, & P. Marris (Eds.), *Attachment across the life cycle* (pp. 127–159). London: Tavistock/Routledge.

Mithaug, D. E. (1993). *Self-regulation theory: How optimal adjustment maximizes gain.* Westport, CT: Praeger.

Moely, B. E., Santulli, K. A., & Obach, M. S. (1993, June). *Strategy instruction, metacognition, and motivation in the elementary school classroom.* Paper presented at a conference on Research on Memory Development: State-of-the-Art and Future Directions. Castle Ringberg, Germany.

Moss, E., Gosselin, C., Parent, S., & Dubeau, D. (1993, March). *Does attachment influence shared metacognitive experiences*? Paper presented at the meetings of the Society for Research in Child Development, New Orleans, LA.

Nath, P., Borkowski, J. G., Whitman, T. L., & Schellenbach, C. J. (1991). Understanding adolescent parenting: The dimensions and functions of social support. *Family Relations, 40*, 411–420.

Nicholls, J. G. (1989). *Competence and accomplishment: A psychology of achievement motivation.* Cambridge, MA: Harvard University Press.

Parent, S., Moss, E. Gosselin, C., & St. Laurent, D. (1993, March). *Maternal adjustment to developing child competencies: Is age all that matters*? Paper presented at the meetings of the Society for Research in Child Development, New Orleans, LA.

Plomin, R., & Bergeman, C. (1991). The nature of nurture: Genetic influence on environmental measures. *Behavioral and Brain Sciences, 14*, 373–428.

Reid, M. K., & Borkowski, J. G. (1987). Causal attributions of hyperactive children: Implications for training strategies and self-control. *Journal of Educational Psychology, 79*, 296–307.

Rodriguez, M. L., Mischel, W., & Shoda, Y. (1989). Cognitive person variables in the delay of gratification of older children at risk. *Journal of Personality and Social Psychology, 52*, 358–387.

Sroufe, A., & Egeland, B. (1991). Person and environment: Illustrations of interactions from a longitudinal study of development. In T. D. Wachs & R. Plomin (Eds.),

Conceptualization and measurement of organism environment interaction (pp. 68–84). Washington, DC: American Psychological Association.

Sternberg, R. J. (1985). *Beyond IO: A triarchic theory of human intelligence*. New York: Cambridge University Press.

van Ijzendoorn, M. H., & van Vliet Visser, S. (1988). The relationship between quality of attachment in infancy and IQ in kindergarten. *Journal of Genetic Psychology, 149*, 23–28.

Wachs, T. D. (1992). *Individual differences and development: The nature of nuture*. Newbury Park, CA: Sage.

Chapter 2

What Environmental Factors Affect Intelligence: The Relevance of IQ Gains Over Time

James R. Flynn

Department of Political Studies, University of Otago, Dunedin, New Zealand

Each generation outscores the preceding generation on IQ tests, often by huge margins. The evidence suggests that this is true for every nation that has begun to transform its society by industrialization. We now have data for 20 such nations and there is not a single exception. They include the advanced nations of Western Europe and virtually all English-speaking nations. The examples of urban Brazil and Israel tempt us to include all nations of European culture, and the examples of urban China and Japan tempt us to include all those that have adopted European technology. Recent data show that IQ gains in Britain began no later than the last decade of the 19th century at a time when, paradoxically, IQ tests did not exist. The time between the advent of industrialization and the beginning of IQ gains is probably short and the two may well coincide (Flynn, 1987, 1994; Raven, Raven, & Court, 1993).

Nations have differed in terms of rate of gain at least since the Second World War. However, superimposed on national differences is a definite pattern, namely that gains are largest on tests of fluid intelligence and tend to diminish as the balance swings towards measuring crystallized intelligence. Gains on culture-reduced tests range from 10 to 20 points (SD = 15) per generation (30 years), with Belgium, The Netherlands, and Israel appearing at the top of the scale, the Scandinavian countries at the bottom, and English-speaking nations in-between. Wechsler performance scale data show a similar magnitude of gain, but national differences cannot be taken as seriously because the data are weaker and rarely include adult subjects. Gains on verbal tests vary from almost nil to 20 points per generation and among 11 nations that allow a comparison, there is not one in which verbal match nonverbal gains. Where vocabulary gains can be measured, they are usually lower still. British adults of all ages, for example, gained 27 points over 50 years on Raven's Progressive Matrices (a nonverbal test of fluid intelligence); but they gained only 6 points over 45 years on the Mill Hill Vocabulary Scale (Emanuelsson & Svensson, 1990; Flynn, 1987, pp. 185–186; 1990, p. 47; Goldenberg, 1991; Lynn, 1990, p. 139; Raven et al., 1993, pp. G22–G26; Teasdale & Owen, 1989).

As for determining what environmental factors affect intelligence, IQ gains over time are relevant in a variety of ways. They have already paid dividends by showing that certain environmental factors lack the potency claimed for them and conversely, that the potency of other environmental factors has been overlooked. They promise additional dividends if only we can begin to understand the cause of IQ gains. Solving the causal problem would, I shall argue, give us a more complete list of factors that raise IQ without raising intelligence. It might even provide us with better measures of intelligence, so we could be confident that factors that raise test scores really do cause intelligence gains.

REASSESSING ENVIRONMENTAL FACTORS

IQ gains over time mean that if you administer a test standardized 20 or 30 years ago, you are scoring subjects against obsolete norms. In other words, the subjects' test performance is being measured not against that of their own generation, but against the poorer performance of the last generation. This inflates their scores in the sense that they appear superior to their contemporaries when they actually are not. Inflated scores have played havoc with assessing the impact of environmental factors on the interrelated traits of intelligence and achievement.

Vernon (1982) relied on studies that inflated the IQ scores of Chinese Americans by scoring them against obsolete norms and mistakenly concluded that they had a higher mean IQ than white Americans. Therefore, he assessed environmental differences between the two groups primarily as to whether these could explain both a Chinese intelligence and achievement advantage. There seemed to be no special problem in the fact that Chinese Americans outperformed white Americans academically and in terms of occupational status by huge margins. As Arthur Jensen put it, the reason Asian Americans do so well may simply be that they are smarter (Brand, 1987, pp. 44–45).

However, when Flynn (1991) adjusted the IQs of Chinese Americans for obsolescence, he found they had no higher mean IQ than white Americans. This posed an entirely new problem: what environmental differences between the two groups could explain the fact that Chinese Americans outperform white Americans without an IQ advantage. Flynn put the Chinese-American IQ/achievement gap at 21 points—that is, Chinese Americans achieve as if they were a group with a mean IQ that is 21 points higher than it actually is. He showed that one-third of this gap could be explained by the fact that Chinese Americans can spot whites 7 IQ points and still match them for academic achievement, which meant that they had a lower IQ threshold for entry into elite occupations. The other two-thirds was explained by a higher capitalization rate—that is, 78% of those who fell above the Chinese threshold for elite professions actually entered those professions, while the figure for whites was only 60%. He hypothesized that environmental factors such as work ethic, the child's perception that the parents genuinely valued him or her because of educational attainment, investment of self-esteem in academic and professional achievement, discipline, and sobriety were the underlying causal variables. The investment of self-esteem in achievement explains the fact that Chinese Americans capitalize more efficiently on their available pool of talent. An Irish-American youth may forfeit a promising opportunity so as to attend a Catholic college, stay with kin or friends, or marry the girl or boy next door—a Chinese youth rarely.

Obsolete norms have tended to obscure the full potency of the distinctive features of the Chinese-American home environment: these features confer what amounts to a 21 IQ-point bonus in terms of group achievement. By contrast, obsolete norms have inflated our estimates of the potency of the distinctive features which differentiate white upper- and middle-class homes from lower-class homes.

In the Milwaukee Project, Heber and Garber attempted to give black ghetto children the advantages of an upper-class home. Heber and Garber gave black ghetto children what, in my opinion, were the advantages of an upper-class home. Beginning at 3 to 6 months of age, the children spent the day at learning centers, enjoyed constant interaction with surrogate parents (paraprofessionals) using the best developmental programs and educational toys known, were provided with the best food and medical and dental care, and their mothers received vocational, home-making, and child-care training.

Heber and Garber believed that they had lifted the mean IQ of these children to over 120—20 points above the average score of contemporary whites. Many at the time, although not Heber and Garber themselves, believed that the intervention had lifted. When corrected for obsolescence, their mean IQ dropped to about 105. Follow-up studies showed an IQ decline after they entered school and that whatever IQ gains they may have enjoyed were not matched by achievement gains (Flynn, 1984b; Jensen, 1989). The famous Skodak and Skeel's Adoption Study found a 20-point gap between the IQ of adopted children and their biological mothers, which seemed to show the potency of being raised in a good adoptive home. However, some of that gap was due to norms that were 14 years out of date, meaning that the adoptive homes were getting credit for what were merely IQ gains over time enjoyed by all children. The childrens' mean IQ was probably no higher than 105.5 and their IQ advantage over their biological mothers somewhere between 10 and 13 points. Whether their eventual achievements were above average is unknown (Flynn, 1993).

It would be wrong to draw firm conclusions simply from ethnic comparisons and adoption studies. However, traits like a person's work ethic, investment of self-esteem in academic and occupational achievement, self-discipline, and sobriety look more important than middle-class affluence, educational toys, and stimulating leisure-time activities. The assets of character are not for sale.

IQ GAINS AND INTELLIGENCE

The causes of IQ gains over time must be environmental—over one or two or even three generations, only a fanatic eugenics program could have made a significant contribution to IQ gains and, if anything, mating trends have been dysgenic. Therefore, it appears that if we knew what causes IQ gains, we would learn much about the environmental factors that affect intelligence. However, this assumes that IQ gains over time can be equated with intelligence gains and that assumption must be questioned.

The brute fact is that at least some gains seem too large to equate with intelligence gains. This assertion presumes an operational concept of intelligence, however rough and ready, and I will label it "understanding-baseball intelligence." It is derived from an account Jensen gave to illustrate the limitations of a subject with a Wechsler IQ of 75. Despite the fact that the man in question volunteered baseball as his chief interest, and attended or viewed games frequently, he was vague about the rules, did not know how many players comprised a team, could not name the teams his home team played, and could not name any of the most famous players (Jensen, 1981, p. 65).

In 1942, J.C. Raven standardized Raven's Progressive Matrices. For ages 20 to 30, he selected soldiers in army camps whose education matched that of British males at the time. For older ages, he tested large samples from a private firm and a government department—a majority of whose employees joined as youths and remained until retirement. They gave him a curve of performance from one age cohort to another and he grafted that curve onto his military sample, thereby deriving norms covering all ages (Foulds & Raven, 1948; Raven, 1941). In 1992, John Raven restandardized Raven's Progressive Matrices on a representative sample of the adult population of Dumfries in Scotland, selected as typical of an area whose norms matched those of Britain as a whole (Raven, 1981, p. RS1.25). The 1992 sample shows that by that year performance on Raven's peaked no earlier than ages 35 to 40. The 1942 sample peaked earlier but maintained top performance until 35 to 40, so those ages allow the fairest comparison. They also match the results of comparing all ages from 18 to 67.

The 1992 sample outscored the 1942 sample by almost 27 IQ points. Moreover, the cohort aged 70 in 1992 matched the 20 year olds from 1942. If we assume that the 70 year olds of 1942 matched the 20 year olds of 1892, we can trace IQ gains back 100 years—that is, back before IQ tests existed. The Raven's data are like rays of light from a distant star which give us a glimpse of the universe as it was long before the earth was born (Raven, et al., 1993, pp. 622–626). The data show that Britons gained 55 IQ points between 1892 and 1992.

This means that in terms of today's norms, adult Britons had an average IQ of 45 in 1892. The samples do not match the quality of military samples and the equating of age cohorts from present to past can only be approximate. Nonetheless it will be difficult to defend any estimate that puts mean IQ in 1892 above 60: therefore, at a minimum, 84% had an IQ below 75. In order to identify IQ gains with understanding-baseball intelligence, we would have to assume that 84% of Britons could not, even if it became their chief interest, understand cricket in 1892. The best data available pose similar problems. Using military data, Dutch males in 1952 had a mean IQ of 80 in terms of 1982 norms. Can we assume that almost 40% of them lacked the capacity to understand their most favored national sports?

These scenarios are derived from gains on tests of fluid intelligence. Jensen's subject had a Wechsler IQ of 75 and Wechsler tests measure a mix of fluid and crystallized intelligence. It may be said that such tests are a better measure of mental retardation and, therefore, our scenarios are suspect. This demands a reply on two levels.

First, there are the US data from tests that measure a mix of fluid and crystallized intelligence. Wechsler and Stanford–Binet samples show gains at an average rate of about 0.3 points per year from 1932 to 1989. However, it is highly probable that these gains began no later than 1918. Every study from that era shows large gains, and they are supported by a comparison of performance on the Stanford–Binet by soldiers in 1918 with the standardization sample of 1932. Collectively, the evidence suggests a gain of at least 21 points between 1918 and 1989 (Flynn, 1984b; Lynn & Pagliari, in press; Terman &

Merrill, 1937, p. 50; Yerkes, 1921, pp. 634 and 789). This means that in 1918, when scored against today's norms, Americans had an average IQ of 79 on tests whose crystallized component is at least as great as Wechsler tests. Does that mean that 40 percent of Americans lacked the capacity to understand the basic rules of baseball?

Second, theory posits a functional relationship between fluid and crystallized intelligence such that their problems cannot be compartmentalized—what afflicts one must affect the other. A population whose fluid intelligence tests at 60 or 70 or 80 should not soar much above that for crystallized intelligence. It is quite possible that subjects whose fluid intelligence did not decline until old age should retain the information and vocabulary they acquired earlier, at a time when their fluid intelligence was normal. The evidence of many studies suggests this (Horn, 1989). However, it is quite another thing to imagine people acquiring normal levels of knowledge and vocabulary whose fluid intelligence never, during their entire lives, rose much above the level of mental retardation. Indeed, the very fact that massive IQ gains can occur, whether measured by tests of fluid or crystallized intelligence, despite nil or minimal vocabulary gains seems incongruous.

Theory also gives tests of fluid intelligence an indispensable role in drawing the distinction between intelligence and learning. As an approach to a pure test of learning, imagine a simple task such as tying one's shoes: Since almost everyone can perform this task, it comes close to measuring no intelligence differences. The Wechsler tests measure a mix of learning and intelligence, that is, the tasks are of graduated difficulty so that they rank people in terms of the amount of intelligence usually required to learn them. Tests of fluid intelligence are designed to presuppose a minimum of learning, often no more than acquaintance with certain simple shapes, and therefore are deemed approaches to a pure test of intelligence (Horn, 1989; Jensen, 1979, 1980).

Jensen and others are now experimenting with replacing IQ tests—at least when they compare groups with results that seem suspect as intelligence differences—with physiological measures. The hope is that by measuring the electrical response of the cerebral cortex to sights and sounds, how quickly people can react to stimuli, and the time taken for an injection of glucose to reach and be absorbed by the brain, we will find a physiologically-based intelligence test that always gives plausible results, even when used to compare generations over time (Jensen, 1988, 1989).

THE TWO CAUSAL PROBLEMS

Massive IQ gains pose two causal problems, and, therefore, candidates for the role of cause must meet two basic criteria: they must be capable of explaining the gains when confronted with the total array of data; and of causing IQ gains without simultaneously causing intelligence gains. At first, there was a reluctance to concede that massive IQ gains had occurred. When the evidence became overwhelming, there was a natural tendency to turn to 20th century trends of obvious relevance such as increased exposure to IQ tests, the advent of television, and advances in education.

The 20th century saw the invention of IQ tests. Exposure to the tests increases test sophistication, familiarity with their format, confidence about coping with them, and so forth, and test sophistication raises IQ scores without any intelligence gains. However, it cannot qualify as a significant cause. IQ gains antedate the period when testing was common and have persisted into the era when testing, due to its unpopularity, has become less frequent. Even when naive subjects are repeatedly exposed to a variety of tests, IQ scores rise by only 5 or 6 points and the rate of gain reduces sharply after the first few exposures. IQ trends over time show gains as high as 55 points and, in some countries, the rate of gain has risen decade after decade.

Thorndike (1977) compiled Stanford–Binet data which suggested that during the period of 1932 to 1971–72, American children aged 6 and under made greater IQ gains than older children. Therefore, he sought causal factors that were likely to affect preschoolers more than others; such as television in general and educational television in particular. Flynn (1984a) compiled a wider array of data which showed that the atypical gains of young children were either an artifact of sampling error or totally antedated 1947, ruling out television as an age-specific factor. Moreover, he used the WISC standardization sample to compare American IQ gains from 1932 to 1947–48 with those from 1947–48 to 1972—the periods immediately before and after the introduction of TV. The rates of gain for both periods were roughly equal. Lynn (1987) hypothesized that children who grew up during the Great Depression and the Second World War would have had their IQs depressed. If so, the WISC standardization sample of 1947–48 would have had an atypically poor performance which would deflate our estimate of gains prior to the introduction of television and inflate our estimate for the period thereafter. In other words, TV might have lowered IQ gains, an effect concealed by the depressed performance of the WISC sample. However, there is ample international data which show massive gains for subjects all born after the Second World War and this counts against the depression–World War hypothesis. (Flynn, 1988, 1990; Goldenberg, 1991; Lynn, 1990; Lynn & Pagliari, in press). There is no reason to believe that TV either increased or reduced the rate of IQ gains in America.

Education is a more serious candidate for the role of possible cause. In most countries, larger numbers of people are spending longer periods of their life being schooled and examined on academic subject matter. IQ gains in Denmark appear highly correlated with increased years of schooling and more people attaining higher credentials (Teasdale & Owen, 1987). However, the reverse is true in The Netherlands, where matching across generations to hold educational level constant eliminates only 6.5% of a massive gain (Flynn, 1987). Gains among schoolchildren, comparing, for example, 6th or 12th graders with their counterparts a generation ago, cannot be influenced by years of schooling, in that the number of years are, by definition, the same. As for quality of schooling, the tendency of IQ gains to escalate the farther one gets from tests that measure traditional academic skills is virtually universal. The United States has had significant IQ gains since 1972, a period during which the National Assessment of Educational Progress (the nation's "report card") found little or no academic achievement gains.

The fact that education cannot explain IQ gains as an international phenomenon does not, of course, disqualify it as a dominant cause at a certain place and time. Particular countries are sometimes influenced by a factor that is culture-specific. Comparing age cohorts suggests that urban China gained 22 IQ points on Raven's Progressive Matrices between 1936 and 1986 (Raven & Court, 1989, p. RS4.8). Learning to read Chinese characters involves endless practice in breaking a complex pattern into two components, one conveying meaning, the other pronunciation. The spread of literacy might be a dominant cause of matrices gains that is peculiar to urban China.

It is logically possible that peculiar factors dominate in each and every one of our twenty countries. But this seems highly unlikely. Two attempts to explain IQ gains as an international phenomenon have gained currency; namely, the Brand hypothesis and the nutrition hypothesis.

Brand argues that the permissive society advantages the present generation on tests with time limits. The scrupulous test taker of the past wasted time trying to get every item correct; today's subjects are prone to intelligent guessing and finish more items within the time allotted. This hypothesis is theoretically ideal. It explains IQ gains in terms of something which implies no intelligence gains and cites environmental factors independent of mere exposure to tests. It has now been proven false, however. Wechsler performance tests in nation after nation show gains as high as the culture-reduced tests, even though the latter usually impose much greater time pressure (Flynn, 1987, pp. 185–186). The huge gains of adult Britons on Raven's were made by subjects who took the test untimed. Flieller, Jautz, and Kop (1989, pp. 11–12) analyzed a Binet-type test with a fairly even balance of timed and untimed items. They found that the last generation left more questions unanswered on both kinds of items; and that poorer performance on items completed accounted for virtually all of the last generation's score deficit.

Unlike Brand's hypothesis, the nutrition hypothesis creates a tension between our two basic causal problems. Better nourished brains would function better in the test room, but they should also function better in everyday life. Therefore, if nutrition has caused 20 or 30 or 50 points of IQ gains, we seem driven to posit huge understanding-baseball intelligence gains as well.

Lynn (1987, 1989) enhanced its plausibility by ascribing only 15 points to nutrition and the remainder to other causes such as defective tests. For example, Raven's Progressive Matrices is held to measure increased arithmetical skills as well as intelligence gains and therefore to overestimate intelligence gains. The critique of Raven's poses many evidential problems. Norwegian draftees made Matrices gains while suffering losses on a test modeled on the Wechsler adult arithmetic subtest. Military samples from Israel show comparable male and female performance on the Matrices, which runs counter to most gender data on mathematics. As for the magnitude of Matrices gains, they are larger than those of other nonverbal tests in Britain, but equivalent in Belgium and smaller in Australia, Canada, and Scotland (Flynn, 1987, 1990; Goldenberg, 1991).

Even if given a diminished explanatory role, the nutrition hypothesis has its own peculiar evidential problems. Lynn (1987, p. 467) cited a one standard deviation height gain over the last fifty years, which equals his estimate of British intelligence gains over that period. However, some European countries have been making height gains for fully a century or two and these amount to more than one *SD*, sometimes to two or three (Floud, Wachter, & Gregory, 1990, pp. 16, 23, and 26). If height gains are truly accompanied by intelligence gains, they pose a familiar question: Did the Dutch in 1864 really have the same intelligence as people who today score 65 on IQ tests? Did Norwegians in 1761 really resemble those who today score 62?

The best experimental study of the effects of vitamin–mineral supplements on IQ shows that in California a modest supplement had little effect, a moderate one had a significant effect, and a large one had little effect (Schoenthaler, Amos, Eysenck, Peritz, & Yudkin, 1991, pp. 357–358). That every nation has continuously enhanced nutrition just the right amount, neither too little nor too much, for decade after decade, seems unlikely. For example, The Netherlands almost certainly gave children born after the Second World War better nutrition than it gave those born during the great war-time famine. The effect on IQ gains over time was nonexistent (Flynn, 1992, p. 346).

The experimental data from dietary supplements also show that 75% of subjects enjoy very modest gains, while 25%, presumably subjects who are subclinically malnourished, make large gains. The latter tend to have lower IQs than the former, which means that if enhanced nutrition is a factor, IQ gains over time should come disproportionately from those with below-average IQs. Denmark fits that pattern, but most nations do not. A good sign that IQ gains extend to every IQ level is that score variance remains unchanged over time, or diminishes only because of clear ceiling effects. Military samples or samples of equivalent quality show this for Belgium, Norway, Sweden, Israel (in males), Canada, and New Zealand. Dutch Raven's data and US Wechsler data actually provide the full IQ curves and allow us to trace gains at all levels (Bouvier, 1969, pp. 4–5; Clarke, Nyberg, & Worth, 1978, p. 130; Elley, 1969, p. 145; Emanuelsson & Svensson, 1990; Flynn, 1985, p. 240; Goldenberg, 1991; Rist, 1982, p. 47; Teasdale & Owen, 1989, pp. 258–259; Vroon, 1984).

Storfer (1990) made an attempt at causal explanation that has much in common with Lynn although he adds an unusual twist at the end. Once again, only a portion of IQ gains are to be identified with intelligence gains—this time 22 points (rather than 15). Once again, the remainder are ascribed to defective tests, this time Raven's is held to be a spatial test measuring a peripheral rather than a core component of intelligence. This new critique of Raven's is equally suspect. Jensen (1980, pp. 646–647) called the notion that Raven's measures perceptual or spatial ability a "common misconception." He emphasises that factorially it measures fluid *g* and little else. Storfer (1990) cast doubt on the concept of "so-called" fluid intelligence and argued that only tests heavily weighted towards crystallized intelligence are true measures of intelligence. The theoretical price to be paid for downgrading fluid intelligence has already been discussed. The plausibility of even a 22-point intelligence gain is not directly confronted. That is, rather than discussing the consequences of putting the mean IQ of our grandparents at 78 in terms of today's norms, the usual within-generation evidence for the validity of IQ tests is cited.

Storfer (1990) supplemented improved nutrition as a cause with factors like the eradication of childhood diseases and improvements in the cognitive quality of the preschool home environment. He argued that these factors, when treated as purely environmental variables, could explain an 11-point intelligence gain since 1900. The analysis takes within-generation data and applies them across generations by making certain assumptions, such as that half of American infants were in unfavorable home environments in 1900, compared to only 20% today. Even positing that these within-generation factors retain their potency when applied between generations, and SES provides an example of a factor whose potency is lost, the analysis adds nothing to our knowledge of the effect of environmental variables on intelligence. It merely makes an unevidenced assumption about the explanatory power of the within-generation data we already have.

The 11 points supposedly explained fall well short of the 22-point intelligence gain over time that Storfer posited. Therefore, he had to double the explanatory potential of his factors by formulating a new Lamarkian theory of evolution and cited cholinergic neurons as the vehicle for the inheritance of an acquired characteristic. These neurons might convey an environmentally-induced change in the brain cells to the testes, allowing that change to be passed from one generation to the next. There is no harm in this sort of speculation, but we await actual anatomical and biochemical evidence. Until then, there is no substantial body of evidence to assess.

HOPES FOR THE FUTURE

Up to now, our efforts to identify the environmental factors that have caused massive IQ gains over time have not come to much. The tendency has been to take the putative 55 IQ-point gain since last century, carve out a small portion to be treated as an intelligence gain, explain that portion by familiar within-generation factors, and treat the remainder as a non-intelligence gain caused by faulty tests. The tests castigated are, it is worth noting, tests hitherto considered reliable and central to the theory of intelligence.

I believe that the portion of IQ gains over time which represents an intelligence gain is very small indeed; and that, paradoxically, the exciting thing to explain is the huge non-intelligence gain. If we could do that, we would win two glittering prizes. At present, our list of factors that cause IQ differences which are not intelligence differences, and therefore likely to deceive us, is very short, running not much beyond test sophistication and test technique. Somewhere out there, however, there are environmental variables of enormous potency in terms of their effect on IQ, which are no more productive of intelligence differences than test sophistication. Some day we may be able to identify them and add them to our list of ersatz causes. It seems likely that these variables operate to some degree within generations as well as between generations. Therefore, they would be well worth knowing. Moreover, IQ gains over time have prompted the search for a measure that will not register ersatz intelligence gains, but, rather, will compare generations for intelligence with plausibility. If we can find such a measure, a better measure of intelligence, we can discover the factors that raise *its scores*. As a consequence, our knowledge of what raises intelligence, as distinct from IQ, will take a giant step forward.

REFERENCES

Bouvier, U. (1969). *Evolution des cotes a quelques tests* [Evolution of scores from several tests]. Brussels: Belgian Armed Forces, Center for Research Into Human Traits.

Brand, D. (1987, August 31). The new whiz kids: Why Asian-Americans are doing so well, and what it costs them. *Time*, pp. 42–51.

Clarke, S. C. T., Nyberg, V., & Worth, W. H. (1978). *Technical report on Edmonton Grade III achievement: 1956–1977 comparisons*. Edmonton, Canada; University of Alberta.

Elley, W. B. (1969). Changes in mental ability in New Zealand schoolchildren. *New Zealand Journal of Educational Studies, 4*, 140–155.

Emanuelsson, I., & Svensson, A. (1990). Changes in intelligence over a quarter of a century. *Scandinavian Journal of Educational Research, 34*, 171–187.

Flieller, A., Jautz, M., & Kop, J-L. (1989). Les réponses au test Mosaïque à quarante ans d'intervalle [Answers on the Mosaique test after an interval of forty years]. *Enfance, 42*, 7–22.

Floud, R., Wachter, K., & Gregory, A. (1990). *Height, health and history*. Cambridge, U. K.: Cambridge University Press.

Flynn, J. R. (1984a). IQ gains and the Binet decrements. *Journal of Educational Measurement, 21*, 283–290.

Flynn, J. R. (1984b). The mean IQ of Americans: Massive gains 1932 to 1978. *Psychological Bulletin, 95*, 29–51.

Flynn, J. R. (1985). Wechsler intelligence tests: Do we really have a criterion of mental retardation? *American Journal of Mental Deficiency, 90*, 236–244.

Flynn, J. R. (1987). Massive IQ gains in 14 nations: What IQ tests really measure. *Psychological Bulletin, 101*, 171–191.

Flynn, J. R. (1988). Japanese intelligence simply fades away: A rejoinder to Lynn. *The Psychologist, 9*, 348–350.

Flynn, J. R. (1990). Massive IQ gains on the Scottish WISC: Evidence against Brand et al.'s hypothesis. *The Irish Journal of Psychology, 11*, 41–51.

Flynn, J. R. (1991). *Asian Americans: Achievement beyond IQ*. Hillsdale, NJ: Lawrence Erlbaum.

Flynn, J. R. (1992). Cultural distance and the limitations of IQ. In J. Lynch, C. Modgil, & S. Modgil (Eds.), *Education for cultural diversity: Convergence and divergence* (pp. 343–360). London: Falmer Press.

Flynn, J. R. (1993). Skodak and Skeels: The inflated mother–child IQ gap. *Intelligence, 17*, 557–561.

Flynn, J. R. (1994). IQ gains over time. In R. J. Sternberg (Ed.), *The encyclopedia of human intelligence* (pp. 617–623). New York: Macmillan.

Foulds, G. A., & Raven, J. C. (1948). Normal changes in the mental abilities of adults as age advances. *Journal of Mental Science, 94*, 133–142.

Goldenberg, J. (1991, March 4). Information about intelligence test scores in Israel. Personal communication.

Horn, J. L. (1989). Cognitive diversity: A framework for learning. In P. L. Ackerman, R. J. Sternberg, & R. Glaser (Eds.), *Learning and individual differences: Advances in theory and research* (pp. 61–114). New York: Freeman.

Jensen, A. R. (1979). The nature of intelligence and its relation to learning. *Journal of Research and Development in Education, 12*, 79–95.

Jensen, A. R. (1980). *Bias in mental testing*. London: Methuen.

Jensen, A. R. (1981). *Straight talk about mental tests*. New York: The Free Press.

Jensen, A. R. (1988). *The Oddman-Out Test: A chronometric anchor for psychometric tests of g*. Unpublished paper.

Jensen, A. R. (1989). Raising IQ without increasing *g*? A review of *The Milwaukee Project: Preventing mental retardation in children at risk. Developmental Review, 9*, 234–258.

Lynn, R. (1987). Japan: Land of the rising IQ. A reply to Flynn. *Bulletin of the British Psychological Society, 40*, 464–468.

Lynn, R. (1989). Positive correlation between height, head size and IQ: A nutrition theory of the secular increases in intelligence. *British Journal of Educational Psychology, 59*, 372–377.

Lynn, R. (1990). Differential rates of secular increase of five major primary abilities. *Social Biology, 38*, 137–141.

Lynn, R., & Pagliari, C. (in press). The intelligence of American children is still rising. *Journal of Biosocial Science*.

Raven, J. (1981). *Manual for Raven's Progressive Matrices and Mill Hill Vocabulary Scales* (research supplement No. 1). London: H. K. Lewis.

Raven, J., & Court, J. H. (1989). *Manual for Raven's Progressive Matrices and Vocabulary Scales* (research supplement No. 4). London: H. K. Lewis.

Raven, J., Raven, J. C., & Court, J. H. (1993). *Manual for Raven's Progressive Matrices and Vocabulary Scales* (section 1). Oxford: Oxford Psychologists Press.

Raven, J. C. (1941). Standardization of Progressive Matrices. *British Journal of Medical Psychology, 19*, 137–150.

Rist, T. (1982). *Det intellektuelle prestasjonsnivaet I befolkningen sett I lys av den samfunns-messige utviklinga* [The level of the intellectual performance of the population seen in the light of developments in the community]. Oslo: Norwegian Armed Forces Psychology Service.

Schoenthaler, S. J., Amos, S. P., Eysenck, H. J., Peritz, E., & Yudkin, J. (1991). Controlled trial of vitamin–mineral supplementation: Effects on intelligence and performance. *Personality and Individual Differences, 12*, 351–362.

Storfer, M. D. (1990). *Intelligence and giftedness: The contributions of heredity and early environment.* San Francisco: Jossey-Bass.

Teasdale, T. W., & Owen, D. R. (1987). National secular treads in intelligence: A twenty-year cross-sectional study. *Nature, 325*, 119–121.

Teasdale, T. W., & Owen, D. R. (1989). Continued secular increases in intelligence and a stable prevalence of high intelligence levels. *Intelligence, 13*, 255–262.

Terman, L. M. & Merrill, M. A. (1937). *Measuring intelligence*. London: Harrop.

Thorndike, R. L. (1977). Causation of Binet IQ decrements. *Journal of Educational Measurement, 14*, 197–202.

Vernon, P. E. (1982). *The abilities and achievements of Orientals in North America.* New York: Academic Press.

Vroon, P. A. (1984, November 5). Raven's score distribution of Dutch draftees. Personal communication.

Yerkes, R. M. (1921). Psychological examining in the United States Army. *Memoirs of the National Academic of Sciences*, Vol. 15. Washington, DC: U.S. Government Printing Office.

Chapter 3

Habits of Mind: Developmental Diversity in Competence and Coping

Daniel P. Keating

Ontario Institute for Studies in Education and Canadian Institute for Advanced Research

Theories of human intelligence that place the greatest emphasis on the heritability of individual differences stress the importance of evolutionary history (and thus the contextual factors that operate through natural selection) in shaping the distribution of genetic factors that influence phenotypic (observed) diversity (Eysenck, 1988; Galton, 1892/1962). In contrast, theories of intelligence that focus on the diversity that arises from different life experiences emphasize the contextual contingencies that operate on individuals (Hunt, 1961).

This traditional dichotomy between nature and nurture—that is, inheritance or genetic accounts versus environmental or experiential accounts—has often obstructed understanding. Given the stark contrast between these two views of the world, it is not surprising that the conflict between them has been heated, in all spheres—scientific, political, cultural. What is more disheartening is that the scientific argument has not moved much beyond the political one in many respects, even though the inaccuracy of this bipolar debate has been recognized for some time (Anastasi, 1958).

THE NATURE–NURTURE DICHOTOMY

Much effort has been devoted to the apportioning of variance between these competing factors, a goal that becomes more complex and nuanced as the precision of these estimates increases (Plomin & Thompson, 1988). For the functioning organism, of course, these influences are never dichotomous, but instead are fully integrated during ontogenesis. A good example of this integration arises from current work that demonstrates the key impact of experiential history on the sculpting of fundamental neural, immune, and hormonal patterns (Cynader, Shaw, Prusky, & Van Huizen, 1990; Suomi, 1991). Historically, relatively less effort has gone toward the construction of robust developmental accounts of how these two substantive influences give rise to the observed diversity in human intellectual performance. To do so, researchers of both persuasions need to shift their focus beyond the apportionment of isolated effects, toward the more complicated task of describing the dynamic interaction of multiple influences over the course of human development (Anastasi, 1986; Bronfenbrenner, Lenzenweger, & Ceci, in press; Green, 1992).

Some past confusion can also be traced to the failure to distinguish between contextual factors in ontogenesis that are associated with intellectual development in general versus those that are associated with the observed diversity in intellectual accomplishment. A useful distinction can be drawn here between capacities and capabilities, a distinction that is masked by the omnibus term "mental abilities." Literacy offers a helpful example. It is obvious that the vast majority of humans have the capacity to become literate, given the appropriate experiential contingencies. Previously illiterate populations demonstrate high proportions of literacy with the advent of schooling, rapidly becoming capable of reading.

Due to theoretical assumptions prevailing in the early history of empirical research on human intelligence (for example, Terman, 1916), these two constructs—capacity and capability—were conflated, in the belief that attained capabilities were quite reliable estimates of fundamental intellectual capacity (Keating, 1990a). More recent efforts to disentangle these notions, in order to achieve purer, more reliable estimates of fundamental capacity *independent* of ontogenetic influences—such as information-processing capacity or neural efficiency—have generated mixed results (Ceci, 1990; Hunt, 1978; Sternberg, 1990). We have argued that the case for "pure" information-processing parameters has yet to be made successfully (Keating, List, & Merriman, 1985; Keating & MacLean, 1987). The natural processes of developmental integration make it difficult to disentangle these influences, stressing again the need to examine in much greater detail the nature of human diversity as a developmental phenomenon (Gardner, 1983; Keating, in press a, in press b).

In summary, the debate in this area has been stuck in a reverberating cycle between positions labeled as "nature" versus "nurture." Given our increasing understanding of the fundamental inseparability of these two broad categories—that is, organisms have built-in structures arising from their phylogenetic history, and ontogenesis always occurs in a physical and social environment that impacts on development—the continuing devolution of an important question into an either/or decision likely reflects the presence of hidden barriers to progress in our understanding. Some of the barriers that impede our understanding of human diversity in intellectual functioning arise from overly simple ideologies and methodologies. Theoretical advances in our understanding of developmental processes, and methodological progress in our ability to study such processes empirically, create the opportunity to move these questions beyond the traditional dichotomies. First, however, a brief review of what we have learned using the traditional models is in order.

Among the best established empirical findings in the behavioral and social sciences are the robust covariance structures of performance on a wide range of cognitive and intellectual tasks (Keating, 1990b; Keating & MacLean, 1988). We may usefully remind ourselves how pervasive these covariance patterns are. In a sufficiently heterogeneous population, positive correlations across a wide variety of cognitive tasks are virtually assured. As well, mean increases in performance with age during the childhood and adolescent years are observed on virtually all cognitive tasks (Case, 1992).

In addition to these robust patterns of individual and age covariance, there are of course stable group differences on many measures of cognitive performance. Patterns associated with demographic indicators, including social class, ethnicity, and gender, have been regularly reported. Organic trauma or genetic anomalies, such as brain lesion or Down syndrome (Cicchetti & Beeghly, 1990), are also reliably related to cognitive performance differences. Patterns of individual differences have also been grouped into diagnostic categories, such as learning disability.

What are the sources of this diversity? The regularity of the patterns described above has contributed to a presumption—among some scientists and the public at large—that a simple and overarching design principle must be responsible. Often, this presumption takes the form of a belief in underlying organismic differences (such as neural efficiency, capacity, or power) as the fundamental source of observed differences.

We should hesitate to make this inferential leap, even though it seems to some more like a short step, appearing to enjoy the benefit of parsimony as well. Empirical efforts to isolate cognitive-processing variance from knowledge-base variance, and vice versa, have encountered substantial methodological obstacles, and, as already noted, the evidence for uncontaminated measures of either hypothetical source remains unpersuasive (Keating & Crane, 1990; Keating et al., 1985; Keating & MacLean, 1987; Morrison, Morrison, & Keating, 1992). Covariance patterns alone, no matter how robust, are insufficient to demonstrate the operation of any specific mechanism.

This example illustrates the confusion that often arises between data structures and inferred mental structures. The two are not the same. Presumably, robust patterns in cognitive performance data must reflect some coherent source, but this does not necessarily imply that there is an organismic structure homologous with any particular data structure. Borrowing from evolutionary logic, we observe the surface similarity but structural dissimilarity between, say, bats and birds, and the surface dissimilarity but structural homology between spiders and crabs. To get at homologies, we need to uncover the underlying processes and their histories (Keating, 1990b). This entails one further understanding: that the impact of experience is also constrained by internal structures which are shaped by both phylogeny and ontogeny. In other words, humans are not infinitely plastic, and remedial experience is not always fully effective.

This historical conflict between dichotomous positions has unnecessarily constrained theory, practice, and research methods that seek to address important questions about human intelligence. We need to know more about human intelligence than how to apportion the amount of influence exerted by two competing categories, each of which is so broad as to be almost wholly uninformative. We need to know *how* human competence and human coping actually develop as self-organizing dynamic systems.

WHAT IS INTELLIGENCE?

The definition of human intelligence has itself been a source of controversy since its inception as a field of research. Broad theoretical definitions—like "general adaptibility"—have proved too difficult to operationalize, and the field has proceeded largely by assuming the operational definition—what the tests measure—as the dominion of the theory. "But after nearly a century, the old gibe is still true: Intelligence is what intelligence tests measure" (Green, 1992, p. 331).

This working assumption has proved to be highly productive in terms of the research it has generated and the insights it has provided. But it should be recognized that it has imposed severe limitations on theory, in that skills or competencies that lie outside the operational definition are excluded from consideration (Gardner, 1983; Keating, 1984, 1990b).

The counterargument to this concern is that new tasks or tests, as they are created, continue to fall neatly within the simple covariance structures that we have replicated for decades. It thus seems reasonable to assume that most tests or tasks that could be developed would also be found to correlate positively with existing measures. The breadth of the domain of general intelligence, using the operational definition, seems adequate to cover most of what might plausibly fall into the domain.

The central weakness of the counterargument, of course, is that the available technology for designing tests or tasks tends to restrict us to methodologies that are themselves consistent with existing approaches. As well, apportioning sources of covariance within the positive correlation matrix, between construct and method covariance, is a daunting task. The proportion of covariance due to similarities among standard assessment methods, however, is far from negligible under any plausible scenario. Furthermore, skills and competencies that have proved intractable in measurement terms are typically placed outside the operational definition, and hence beyond the theoretical reach of an empirical science.

The case of social intelligence is an instructive one. Recently, this area has enjoyed renewed interest from several diverse perspectives. Important links between social connectedness and immune system functions have been reported in psychoneuroimmunology, with significant consequences for physical and mental health (Kiecolt-Glaser & Glaser, 1991). The ability to function in groups, where collaboration is essential for success, may be increasingly important in educational, scientific, and workplace contexts (Resnick, Levine, & Teasley, 1991), as it has been important in human history for economic and material advancement (Stringer & Gamble, 1993; Rosenberg, 1986).

Our experimental studies on social intelligence and social cognition converge with the conclusions obtained by other researchers. It is difficult to obtain any construct-valid differentiation between academic-type intelligence measures and social intelligence measures. This seems to be true for many social cognitive measures, including detection of nonverbal communication (Keating & MacLean, 1991); several kinds of social problem solving (Keating, 1978; Keating & Matthews, 1990); and grasp of interpersonal relationships (Keating & Clark, 1980). In each of these cases, the parsimonious interpretation is that variance on these tasks is substantially intercorrelated with most other traditional cognitive and intellectual tasks, seeming to support the view that we have already "captured" the essentials in traditional general intelligence theories.

There is one major empirical difficulty with this parsimonious interpretation. Success on tasks like these appears to be almost wholly unrelated to social competence, if we define social competence as doing well in real-world social situations, based on criteria of peer, teacher, and self judgments (Ford, 1982; Keating & MacLean, 1991; Keating & Matthews, 1990). If we define social competence by social status (attained education, occupation, income) the correlation with general intelligence re-emerges; but, as criteria of social competence, social status indicators are of course hopelessly confounded with the diversity arising from differences in background—the "demographic addresses" of social class, gender, and cultures (Bronfenbrenner & Crouter, 1983).

One might imagine that similar difficulties plague other hard-to-measure human competencies, such as emotional awareness (Oatley & Jenkins, 1992). Indeed, the relative difficulty of measuring the socio-emotional domain in a way that captures its real diversity, rather than reducing it to easily measurable dimensions, has tended to exclude it from concerns about human intelligence. But this is clearly a theoretical bias, based on assumptions about what aspects of human intelligence are central. Which aspects are indeed central is an interesting and open question, to which a wide range of evidence should be brought to bear, including phylogenetic, sociohistorical, and ontogenetic findings. We might begin with the current claims that social interaction, particularly social talk, is the nexus for the evolution of intelligence in Homo sapiens (Dunbar, 1992; Stringer & Gamble, 1993; Tomasello, Kruger, & Ratner, 1993).

But the *a priori* limitation on the theoretical breadth of the domain of human intelligence to those features that we are currently able to measure is surely nonscientific. It is analogous to suggesting that theoretical discussion of the origin of the universe should be excluded from cosmology because we do not yet have equipment that enables us to see back that far.

BEYOND TRADITIONAL DICHOTOMIES: CREATING A HUMAN DEVELOPMENTAL METHODOLOGY

The inherited ability and the environmentalist positions have been interpreted as though the main effects were the ones that truly mattered. But both interpretations ignore the central reality: the only truly causal pathways are embedded in the transactions between the organism and the environment over time.

> Both [correlational and experimental] methods favor main effects over interactions, whereas the development of all aspects of life, including intelligence, involves an interaction of heredity with environment. But dissecting that interaction will require a level of detail and precision not now available. (Green, 1992, p. 331)

One problem in pursuing this research agenda is thus its seemingly overwhelming complexity. The number of potentially important factors and their interactions expands exponentially as we take into account higher-order interactions, and even more so if we examine the multiplicative interactions of those factors across time. Beginning with even a small set of factors, we quickly approach an effectively infinite set of possible causal arrangements, at least some of which will be indistinguishable on the basis of statistical fit to a model (Glymour, Scheines, Spirtes, & Kelly, 1987).

Can we hope to deal with this level of complexity? We must learn to do so, if we hope to address pressing theoretical and practical questions. Specifically, we need to create a developmental methodology for integrating analyses from many sources in a robust fashion (Gould, 1986; Keating, in press c).

Paradoxically, dissecting the interaction should be as interesting to pure nature as to pure nurture types. The link between genetic and intellectual variability, to be viewed as causal, requires a developmental account of how that genetic variation is translated into behavior. In other words, available behavior-genetic evidence speaks to the fact *that* genetic variation is implicated in behavioral variation, not with *how* that connection is established.

One possible explanation is that genetic variability codes directly for neural efficiency in some way. Given the fascinating evidence now emerging on neural sculpting and reorganization that occurs as a function of the transaction of the organism with the environment, direct coding for size, capacity, or efficiency is an assumption that requires re-examination (e.g., Cynader et al., 1990; Turner & Greenough, 1985). It is also worth noting that heritability varies with the age of the group in which it is examined, suggesting that the developmental routes of genetic expression are unlikely to preserve one-to-one correspondence with particular phenotypic characteristics (Plomin & Thompson, 1988).

Another behavior-genetic candidate is temperament, which does show stable variation very early in infancy. But again, the temperament of the infant interacts with characteristics of the primary caregiver (usually the mother), so that a one-to-one correspondence between infant temperament and later behavioral outcome seems implausible. Genetically controlled experimental studies of rhesus macaques show this interaction quite clearly, in that genetically hyper-reactive (and thus highly vulnerable) infant monkeys cross-fostered to highly nurturant mothers are more likely than normal infants to become group leaders (Suomi, 1991).

To advance our understanding of human intelligence, then, we will need to examine in greater detail the history of interactions between the organism and the environment, whatever our theoretical predilections. In this investigation, our evolutionary heritage, especially our primate histories, will be of prime relevance. Numerous examples come quickly to mind. The socially shared nature of much cognitive activity, of increasing practical importance (Resnick et al., 1991), implicates interpersonal relationships and social competence. Our primate history as a social species thus plays an important role in competence and productivity (Suomi, 1991). As well, the major role that emotional aspects play in cognition has been emphasized in recent work (Oatley & Jenkins, 1992), which in turn invoke psychoneuroimmune links (Kiecolt-Glaser & Glaser, 1991).

Integrating human development so as to encompass all these histories—evolutionary, cultural, ontogenetic—requires that many specific methods be used. Each of these methods needs to be critically examined for its ability to permit robust inferences, but we should not assume that we can define robustness with paradigmatic criteria.

Progress in methodology has changed the way we can investigate human intelligence (Keating, in press c). To reflect this shift, I often use the term developmental diversity as a more inclusive one than individual differences, which implies a primary origin within the individual, rather than a historical transaction between the person and a cognitively socializing habitat (or bio-ecology—Ceci, 1990).

DEVELOPMENTAL INTEGRATION OF HUMAN INTELLIGENCE

What shape might a developmental integration on human intelligence take? Recognizing that history is always contingent, contextualists never predict the future with certainty. But odds now favor efforts to weave these strands into a coherent story of human development, in all its diversity. Our attempt to tell a developmentally integrated story (Keating, 1990a, 1990b; Keating & MacLean, 1988) begins with the premise that human intelligence is a dynamic system, at two major levels—populations and individuals.

It grows increasingly clear that knowledge of all types—practical, scientific, theoretical—is always a social and cultural product. As advanced information technologies spread, the social nature of knowledge will become ever more apparent. The emerging picture of science as a collaborative, cumulative discourse captures the essence of one key self-organizing social system. Differences among societies in how well they are able to make use of the social nature of knowledge may determine, in part, how effective they will be in building successful, innovation-based economies. In other words, socially distributed intelligence may become increasingly central to societal success. It depends in turn on the diversity of talent available in the population *and* on the ways in which human groups interact to become units of learning.

Human intelligence is also a self-organizing system at the individual level. A handful of critical elements are essential to the story (Keating, 1990b). A first principle of dynamic systems—behavioral, biological, or physical—is that they display the capacity to become organized over time, even from ill-formed or chaotic origins. An important corollary is that infinitely elaborate and formally elegant structures can arise from simple feedback processes operating over time. This arises from four central features of dynamic systems, two of which describe functions and two others describe the operational context: a) the process must iterate routinely; b) the process must have a feedback loop; c) the context includes internal constraints that shaped the system—for organisms, their phylogenetic and the ontogenetic history to date; and d) the context sets the external constraints that limit and shape the actual self-organization that takes place.

These principles can then define a general developmental function. These factors of process and context interact over time, and the history of those interactions is incorporated into organismic structures. As in the evolution of species, an individual's future is never fully determined by the past, but it is always constrained by it. Several consequences follow from the nonlinear nature of dynamic systems. We should expect causes and effects to be multidimensional, and often mutually causal. Also, the magnitude of causes and effects will not always be commensurate. Indeed, the timing of even minor events may lead to a cascade of other events whose outcome for the individual was far greater than the seeming magnitude of its origin.

HABITS OF MIND: DEVELOPMENTAL DIVERSITY IN COMPETENCE AND COPING

One of the crucial discoveries in recent studies of cognitive development is the fundamental non-independence of cognitive activities from their content and context. An equivalent, but more positive, term for this non-independence is connected, or perhaps even better, integrated cognitive activity. To locate homologous structures in cognitive activity, we need to study its ontogenesis along with its current functioning. This requires much more than looking at the correlation of age with cognitive performance, since that reflects in large part the averaging effects of cognitive socialization environments. We need to study the dynamic interactions among emerging cognitive structures and the cognitive socialization niches within which they develop (Ceci, 1990; Keating, 1990a; Resnick et al., 1991; Rogoff & Lave, 1984). For some time, we have been exploring the developmental processes, especially in infancy, childhood and adolescence, that underlie competence, both in specific expertise and in general habits of mind.

In seeking to account for the early growth of conceptual knowledge, we need to be aware of Gibsonian phenomenological priors that constrain how we see the world, and how these pre-attuned perceptual patterns are shaped by experiences in infancy into basic conceptual structures (Keating & MacLean, 1988; MacLean & Keating, 1993; MacLean & Schuler, 1989). Intuitive conceptions and misconceptions acquired early in life interact differentially with schooling experiences; these interactions are another potent source of developmental diversity (Gardner, 1991; Keating & Crane, 1990). Understanding these interactions is aided by detailed investigations into the role of automaticity; the organization and content of procedural and declarative knowledge; the function of self-regulating cognitive activities (like metacognitive strategies or control processes); and social, emotional, and motivational factors, all of which influence the development of intelligence. However, instead of the "20-questions" approach ("Does performance on X cognitive activity covary with performance on Y complex task?"—to which the answer is almost always "Yes, a small amount"), we need to study how these various aspects of cognitive activity become coordinated over time as individuals develop expertise and competence (Keating, 1990a; Keating et al., 1985; Keating & MacLean, 1987).

One potential outcome of an integrative approach is the detailed description of pathways to the development of expertise (Keating, 1990a). The long-lived controversy between general and specific theories of intelligence, seen in developmental terms, becomes productive rather than unresolvable (Keating & Crane, 1990). For example, even among adolescents who are in the top 5% of the general intelligence dimension, there are diverse and highly consistent patterns of competence (linguistic, social, technical) that are also strongly associated with "non-cognitive" factors such as goals, future aspirations, out-of-school activities, and perceived self-competence (Keating & Matthews, 1990). On the practical side, knowledge arising from the simultaneous study of development and pedagogy contributes effectively to instruction (Keating, 1991, in press b).

The domain-generality versus domain-specificity debate takes on a new look in light of these considerations. I think our common structural metaphors (various poly-archies or hierarchies, structures of intellect) are somewhat too restrictive for the distinction I intend. I find it useful to identify two kinds of habits of mind: those that relate to domain-specific expertise and those that are more domain-general.

The metaphor of habits of mind has several advantages. First, it presumes no particular structural outcome. Rather than reducing diversity to fit an a priori pattern, it encourages the observation of what fits with what over time. In so doing, it allows appropriate degrees of freedom to the operation of contingent history. Second, it strikes a better balance between the inevitably closed or fixed quality of structures (Keating, 1990c) and the apparent plasticity of development. We know a bit about habits—the longer we have them, the harder they are to change, but they are very flexible in the early stages, and are never completely rigid. Third, habits of mind are not exclusively cognitive. They can easily incorporate dispositional, emotional, motivational, and personality variability as well, which is a clearly desirable goal (Keating & Crane, 1990; Sternberg, 1989). These developments are probably closely linked with other important habits of mind, namely coping skills and orientations (Keating, Menna, & Matthews, 1992; Menna & Keating, 1992). These, too, are not merely cognitive, but also social and emotional.

We may well discover that habits of coping that are most important for health and well-being—maintaining social connectedness and exercising reasonable control over one's choices—are similar to, and perhaps even homologous with, the broad habits of mind that shape the acquisition of competence. It is likely that the most developmentally sensitive period for laying the groundwork of later competence and coping occurs during the infant's earliest social interactions, probably in the first two years of life. Basic habits of mind that guide how we interact with others, how we attend to the world, what we focus our attention on, and how we learn to deal with new situations are shaped in the context of these key social relationships (Keating, 1990b; Lewis, 1993; Tunstall, 1993).

Whether or not this particular account of developmental diversity proves true, it is clear that detailed dissections of critical developmental interactions will yield a far more complete and useful picture of human intelligence than the one-dimensional stories that still dominate too much current thinking. To do so, we will need to go beyond the historical constraints of traditional models. In going beyond the traditional models, however, we need to proceed thoughtfully, both to preserve the valuable knowledge we have already gained, and to prevent unanalyzed methodological and ideological assumptions from distorting our perspective.

In simplest terms, early models of human intelligence focused on the question of who had more or less intellectual capacity, and to what those differences should be attributed. We need to move now toward a developmental model, whose focal questions are different. How are the universal human capacities—for language, social interaction, forward planning, and abstract thinking—translated into attained capabilities? And perhaps most important, how can we arrange the human social environment so as to maximize competence and coping in the population, maintain the valuable diversity of domains of expertise, and create the social frameworks that facilitate productive networks for collaborative learning?

REFERENCES

Anastasi, A. (1958). Heredity, environment, and the question "*How?*" *Psychological Review, 65*, 197–208.

Anastasi, A. (1986). Experiential structuring of psychological traits. *Developmental Review, 6*, 181–202.

Bronfenbrenner, U., & Crouter, A. C. (1983). The evolution of environmental models in developmental research. In P. H. Mussen (Ed.), *Handbook of child psychology* (4th ed., Vol. 1) (pp. 357–414). New York: Wiley.

Bronfenbrenner, U., Lenzenweger, M. F., & Ceci, S. J. (in press). Heredity, environment, and the question: "*How?*" In R. Plomin & J. McClearn (Eds.), *Nature, nurture, and psychology*. Washington, DC: American Psychological Association.

Case, R. (1992). *The mind's staircase: Exploring the conceptual underpinnings of children's thought and knowledge*. Hillsdale, NJ: Erlbaum.

Ceci, S. J. (1990). *On intelligence . . . more or less*. Englewood Cliffs, NJ: Prentice-Hall.

Cicchetti, D., & Beeghly, M. (Eds.). (1990). *Children with Down syndrome: A developmental perspective*. New York: Cambridge University Press.

Cynader, M., Shaw, C., Prusky, G., & Van Huizen, F. (1990). Neural mechanisms underlying modifiability of response properties in developing cat visual cortex. In B. Cohen & I. Bodis-Wolliner (Eds.), *Vision and the brain: The organization of the central visual system.* New York: Raven Press.

Dunbar, R. (1992). Why gossip is good for you. *New Scientist, 136*, 28–31.

Eysenck, H. J. (1988). The concept of "intelligence": Useful or useless? *Intelligence, 12*, 1–16.

Ford, M. E. (1982). Social cognition and social competence in adolescence. *Developmental Psychology, 18*, 323–340.

Galton, F. (1962). *Hereditary genius: An inquiry into its laws and consequences.* Cleveland, OH: World Publishing Company. (Original work published 1892)

Gardner, H. (1983). *Frames of mind: The theory of multiple intelligences.* New York: Basic Books.

Gardner, H. (1991). *The unschooled mind.* New York: Basic Books.

Glymour, C., Scheines, R., Spirtes, P., & Kelly, K. (1987). *Discovering causal structure.* Orlando, FL: Academic Press.

Gould, S. J. (1986). Evolution and the triumph of homology, or why history matters. *American Scientist, 74*, 60–69.

Green, B. F. (1992). Exposé or smear? The Burt affair. *Psychological Science, 3*, 328–331.

Hunt, E. (1978). Mechanics of verbal ability. *Psychological Review, 85*, 109–130.

Hunt, J. McV. (1961). *Intelligence and experience.* New York: Ronald Press.

Keating, D. P. (1978). A search for social intelligence. *Journal of Educational Psychology, 70*, 218–223.

Keating, D. P. (1984). The emperor's new clothes: The "new look" in intelligence research. In R. Sternberg (Ed.), *Advances in the psychology of human intelligence* (Vol. 2) (pp. 1–35). Hillsdale, NJ: Erlbaum.

Keating, D. P. (1990a). Charting pathways to the development of expertise. *Educational Psychologist, 25*, 243–267.

Keating, D. P. (1990b). Developmental processes in the socialization of cognitive structures. In *Entwicklung und Lernen: Beiträge zum Symposium anläβlich des 60. Geburtstages von Wolfgang Edelstein.* [Development and learning: Proceedings of a symposium in honour of Wolfgang Edelstein on his 60th birthday] Berlin: Max Planck Institut für Bildungsforschung.

Keating, D. P. (1990c). Structuralism, deconstruction, reconstruction: The limits of reasoning. In W. F. Overton (Ed.), *Reasoning, necessity, and logic: Developmental perspectives* (pp. 299–319). Hillsdale, NJ: Erlbaum.

Keating, D. P. (1991). Curriculum options for the developmentally advanced: A developmental alternative for gifted education. *Education Exceptionality Canada, 1*, 53–83.

Keating, D. P. (in press a). Contextualist theories of human intelligence. In R. J. Sternberg, S. J. Ceci, J. Horn, E. Hunt, J. D. Matarazzo, and S. Scarr (Eds.), *Encyclopedia of intelligence.* New York: Macmillan.

Keating, D. P. (in press b). Developmental diversity in mathematical competence. In L. Penner, G. M. Batsche, H. M. Knoff, D. L. Nelson, & C. D. Spiegelberger, (Eds.), *Contributions of psychology to mathematics and science education.* Washington, DC: American Psychological Association.

Keating, D. P. (in press c). Understanding human intelligence: Toward a developmental synthesis. In C. Benbow & D. Lubinski (Eds.), *From psychometrics to giftedness: Essays in honor of Julian Stanley*. Baltimore, MD: Johns Hopkins University Press.

Keating, D. P., & Clark, L. V. (1980). Development of physical and social reasoning in adolescence. *Developmental Psychology, 16*, 23–30.

Keating, D. P., & Crane, L. L. (1990). Domain-general and domain-specific processes in proportional reasoning. *Merrill-Palmer Quarterly, 36*, 411–424.

Keating, D. P., List, J. A., & Merriman, W. E. (1985). Cognitive processing and cognitive ability: A multivariate validity investigation. *Intelligence, 9*, 149–170.

Keating, D. P., & MacLean, D. J. (1987). Cognitive processing, cognitive ability, and development: A reconsideration. In P. A. Vernon (Ed.), *Speed of information-processing and intelligence* (pp. 239–270). Norwood, NJ: Ablex.

Keating D. P., & MacLean, D. J. (1988). Reconstruction in cognitive development: A post-structuralist agenda. In P. B. Baltes, D. L. Featherman, & R. M. Lerner (Eds.), *Life-span development and behavior* (Vol. 8) (pp. 283–317). Hillsdale, NJ: Erlbaum.

Keating, D. P., & MacLean, D. J. (1991). *Social competence and social cognition in late adolescence and youth: What is the connection*? Paper presented at the meeting of the Society for Research in Child Development, Seattle, WA.

Keating, D. P., & Matthews, D. (1990). *Domain generality and specificity in higher order thinking*. Paper presented at the meeting of the Society for Research in Adolescence, Atlanta, GA.

Keating, D. P., Menna, R., & Matthews, S. (1992). *Adolescent cognitive development in everyday life*. Paper presented at the meeting of the Society for Research in Adolescence, Washington, DC.

Kiecolt-Glaser, J. K., & Glaser, R. (1991). Stress and immune function in humans. In R. Ader, D. L. Felten, & N. Cohen (Eds.), *Psychoneuroimmunology* 2nd ed. (pp. 849–867). New York: Academic Press.

Lewis, M. D. (1993). Emotion–cognition interactions in early infant development. *Cognition and Emotion, 7*, 145–170.

MacLean, D. J., & Keating, D. P. (1993). *Developmental integration of the object concept*. Unpublished manuscript, Department of Psychology, Brock University, St. Catharines, Ontario.

MacLean, D. J., & Schuler, M. (1989). Conceptual development in infancy. *Child Development, 60*, 1126–1137.

Menna, R., & Keating, D. P. (1992, June). *Coping with emotions in adolescence and early adulthood: Developmental changes*. Paper presented at the meeting of the Canadian Psychological Association, Quebec City, Quebec.

Morrison, G. R., Morrison, S. R., & Keating, D. P. (1992). On estimating processing variance: Commentary and reanalysis of Kail's "Developmental functions for speeds of cognitive processes." *Journal of Experimental Child Psychology, 54*, 288–307.

Oatley, K., & Jenkins, J. M. (1992). Human emotions: Function and dysfunction. *Annual Review of Psychology, 43*, 55–85.

Plomin, R., & Thompson, L. (1988). Life-span developmental behavior genetics. In P. B. Baltes, D. L. Featherman, & R. M. Lerner (Eds.), *Life-span development and behavior* (Vol. 8) (pp. 1–31). Hillsdale, NJ: Erlbaum.

Resnick, L. B., Levine, J. B., & Teasley, S. D. (Eds.). (1991). *Perspectives on socially shared cognition.* Washington, DC: American Psychological Association.

Rogoff, B., & Lave, J. (Eds.). (1984). *Everyday cognition: Its development in social context.* Cambridge, MA: Harvard University Press.

Rosenberg, N. (1986). *How the West grew rich.* New York: Basic Books.

Sternberg, R. J. (1989). Domain-generality versus domain-specificity: The life and impending death of a false dichotomy. *Merrill-Palmer Quarterly, 35*, 115–129.

Sternberg, R. J. (1990). *Metaphors of mind: Conceptions of the nature of intelligence.* New York: Cambridge University Press.

Stringer, C., & Gamble, C. (1993). *In search of the Neanderthals.* New York: Thames and Hudson.

Suomi, S. J. (1991). Early stress and adult emotional reactivity in rhesus monkeys. In *The childhood environment and adult disease.* (Ciba Foundation Symposium 156). Chichester, England: Wiley.

Terman, L. S. (1916). *The measurement of intelligence.* Boston, MA: Houghton-Mifflin.

Tomasello, M., Kruger, A. C., & Ratner, H. H. (1993). Cultural learning. *Behavioral and Brain Sciences, 16*, 495–552.

Tunstall, T. K. (1993). *Mother–child interactions in infancy and at two-and-a-half years as predictors of cognitive competence at four-and-a-half years.* Unpublished doctoral dissertation, University of Toronto, Toronto, Ontario.

Turner, A., & Greenough, W. T. (1985). Differential rearing effects on rat visual cortex synapses. I: Synaptic and neuronal density and synapses per neuron. *Brain Research, 329*, 195–203.

Chapter 4

Developmental Systems Challenges to the Study of Specific Environmental Effects: An Argument for Niche-Level Influences*

Robert C. Pianta
Thomas G. O'Connor

University of Virginia

The hypothesis that specific environmental variables effect specific changes in intelligence greatly influences research on the relation between the environment and IQ (Bornstein, 1989; Bradley, Caldwell, & Rock, 1988; Pianta & Egeland, 1994; Wachs, 1992; Wachs & Gruen, 1982). Although an important

* This research was supported in part by NICHD grant R01HD26911 and NIDRR grant H133G20188. Address requests for reprints to the first author at: 147 Ruffner Hall, 405 Emmet St., University of Virginia, Charlottesville, VA 22903.

response to global characterizations of environmental action, there are at least four serious challenges to the specificity hypothesis that flow from a systems view of development (Sameroff, 1989): a) the environment as a system; b) covariation of influences within and across levels of the environment; c) imprecise and inconsistent measurement of "specific" variables, and d) implications of genetic (biological) models for research on environmental effects (Wachs, 1993). We suggest the concept of developmental niche (Super & Harkness, 1986) as a response to these challenges.

SYSTEMS MODELS OF ENVIRONMENTAL INFLUENCE

It is widely accepted that a systems model is the best approximation of the complex nature of the environment (e.g., Lerner, 1989). However, the pursuit of distinct, "specific" environmental influences runs counter to the assumption that the environment is a *system*. Dynamic, transactional systems models emphasize multidirectional relations from which evidence for specific dimensions (Ex, Ey) must be extracted—however, the idea and process of extraction are somewhat artificial within systems theory (Sameroff, 1983). If the assumption of the environment as a system is accepted, then pursuit of specificity of effect has to be balanced by considerations of the holistic organization that constrains that effect (Ogbu, 1987).

Descriptions of environmental systems vary from complex, multi-level, multi-factor models (Bradley, 1989; Lerner, 1989; Ramey, MacPhee, & Yeates, 1982; Wachs, 1992) to simplified two and three level models (Bronfenbrenner & Ceci, 1993; Sameroff, 1989). Bronfenbrenner and colleagues distinguished direct proximal processes (the "engine" of cognitive growth) from a class of moderators (e.g., social networks, culture) termed "context" variables. Bradley (1989) discussed proximal process in terms of source, mode of transmission, function, reactivity, intensity and complexity—300 possible combinations in an eight-dimension space. Sameroff (1989) differentiated between regular patterns of caretaking behaviors and family roles/functions (mini-regulations), the regulatory action of interactions (micro-regulations), and cultural/contextual influences (macro-regulations). Lerner (1989) posited a model with seven levels, each encompassing multiple agents of influence, all connected bidirectionally.

These models are used to generate and explain relations between measures of the environment and IQ (Pianta & Egeland, 1994; Wachs, 1992). However, the distinctness ("specificity") of many of the influences posited within models has not been substantiated. For example, Wachs and Gruen (1982), described reciprocal interaction (mutual interaction, responsivity, sensitivity), verbal stimulation, parental restrictions of exploratory behavior, and parental teaching, as well as warmth, hostility, and control, as separate aspects of parent–child interaction. However, they all overlap at the behavioral level, as we discuss later. Bakeman, Adamson, and Strisik (1989) suggest a focus on specific, independent constructs assessed using standardized procedures across studies. Furthermore, studies reporting bivariate relations between an environmental parameter and IQ are frequently used to support the specific effects hypothesis without reference to the systems model from which the bivariate relation was derived.

COVARIATION ACROSS LEVELS OF ENVIRONMENTAL INFLUENCE: EFFECTS OF NICHES

Conceptual models of the environment posit multiple levels and multiple agents of influence within levels that are bidirectionally linked. Covariance and interaction among environmental influences are usually addressed by partialling background correlations through statistical and sample selection techniques (Pianta & Egeland, 1994). However, systems theory (Lerner, 1989; Sameroff, 1989; Super & Harkness, 1986) posits a role for the concerted action of the *organization* of environmental influences that has implications for how researchers think about and study environmental influence.

Levels of the social environment are interdependent: Superordinate levels influence subordinate levels (Sameroff, 1989) by regulating tolerances or dictating the timing of regulatory events (Ogbu, 1987) while subordinate levels create challenges to superordinate influence, all in dynamic multidirectional exchange (Super & Harkness, 1986). Sameroff used the term "environtype" to denote the organization of environmental influences *across levels* that regulates the behavior of the organism and how it fits within society. Super and Harkness (1986) used the term "developmental niche" to describe similar co-active processes across the multiple levels of a) physical and social settings; b) customs of child care and child rearing, and; c) the psychology of individual caregivers. In this multilevel perspective, proximal process does not have an effect that can be separated from the niche in which it occurs—proximal process reflects a *point of contact* between the child and its developmental niche. Ethnographic studies, multivariate assessments of environmental action, and early intervention studies indicate that specific environmental effects (such as maternal responsiveness) cannot be meaningfully isolated from the larger niche in which they occur (e.g. Marvin & Pianta, 1992; Super & Harkness, 1986).

Multivariate Studies of Environmental Influences

Sameroff, Seifer, Baldwin, and Baldwin (1993) and Sameroff, Siefer, Barocas, Zax, and Greenspan (1987) examined children's IQ in relation to ten social/family risks to which they were exposed including distal (SES, minority status) and proximal (interaction) process. In Sameroff et al. (1993), these ten risk indices were cluster analyzed to describe different packages of risk—a multivariate description of a niche. Clusters differed significantly on child IQ, but when adjusted for number of risk factors the differences disappeared—level, not type of risk, was most related to IQ. In one of several studies, Mink, Blacher, and Nihira (1988) cluster analyzed family social context measures for families with a child with a disability. Seven distinct clusters appeared across a range of SES, ethnicity, and disability conditions. Findings of differential child outcomes associated with cluster membership suggest that, unlike the Sameroff et al. data, different niches may have specific effects on development apart from general level of risk. When multiple *levels* of the environment are included in multivariate analyses, and covariation across levels is included in the description of the environment (as opposed to excluded), it then becomes less tenable to view the impact of the environment in terms of single proximal influences.

Early Intervention Research and "Niche" Effects

Effects of niche-like aspects of the environment are also supported when IQ changes accompany multilevel changes in the environment, as in early intervention programs (e.g., Achenbach, Phares, Howell, Rauh, & Nurcombe, 1990; Breitmayer & Ramey, 1986; Burchinal, Lee, & Ramey, 1989; Lazar & Darlington, 1983; Lee, Brooks-Gunn, Schnur, & Liaw, 1990; Wasik, Ramey, Bryant, & Sparling, 1990). Twenty-five years of implementing early intervention programs demonstrated that targeting a single aspect of the environment for change (e.g., mother–child interaction, play materials) has little effect on child IQ. Multilevel, multiinfluence interventions may have greater effects (Ramey, Bryant, & Suarez, 1985; Wasik, et al, 1990) in part because they involve multiple components of the developmental niche: social and physical settings, parental beliefs about childrearing (e.g., discipline), nutrition, medical care, or parent education/job training. Although the intervention research community has been challenged to produce evidence for the specific aspects of the environment responsible for cognitive change (Wachs, 1992), understanding and describing the effects of the niche is more fruitful than unpacking the effects of its parts.

COVARIATION WITHIN LEVELS OF ENVIRONMENTAL INFLUENCE: MEASUREMENT ISSUES IN ATTRIBUTING SPECIFICITY

Support for the specificity hypothesis requires that researchers distinguish conceptually, operationally, and statistically between "specific environment dimension Ex and specific environment dimension Ey" (Wachs, 1992, p. 77). How

Ex and Ey are measured effects inferences about specificity. Three sources of error plague measurement of *specific* environmental variables: covariation among "specific" variables within the same level of influence, heterogeneity in measurement strategies for the same variable across studies, and imprecise measurement. Literature reviews that find evidence for the effects of specific environmental variables by compiling findings across studies are contaminated with these three sources of error with respect to specificity. We illustrate these problems using the specific construct of maternal responsivity.

With regard to observational studies, Bornstein and Tamis-LeMonda (1989) defined responsiveness as "mothers' prompt, contingent and appropriate behaviors" (p. 50), which is further differentiated into seven modes of maternal behaviors to distress or nondistress infant states. Observations were conducted in the home for up to one hour in both US and Japanese samples. Evidence for specific effects is supported by detailed examination of the responsiveness construct (and its constituent behaviors) according to established taxonomic criteria that address within-level covariation (Bornstein, 1989). Social and didactic responsiveness appear somewhat orthogonal and differentially related to cognitive performance. Although Bornstein's data provide the best evidence for effects of a specific environmental variable, it comes at the expense of neglecting across-level covariation of environmental influences; subjects fall within a narrow band of SES. In another example, Wachs (1984) found evidence for specific effects of contingent responding based on correlations between 60 observational codes composited within each of 4 time periods and 8 Piagetian scales administered on several occasions. Although the evidence for specificity is interpreted as strong on the basis of selected validity coefficients, correlations among the 60 environmental codes (within-level covariance) are not reported. In a recent study (Wachs, Bishry, McCabe, Galal, & Shaheen, 1993), across-level covariation is addressed by using an Egyptian sample; however, findings are again not interpreted with respect to within-level covariation of environmental variables. Finally, Beckwith and Parmelee (1986) observed "caregiver responsiveness" in different forms at different ages using a large number of behaviors coded every 15 seconds. These behaviors were factored at each age, and a first factor named "positive maternal attentiveness" accounted for between 31 and 61 percent of the variance. Despite the high degree of covariation among behavioral indicators of responsiveness, Beckwith and Cohen (1984) reported correlations between individual responsiveness items and cognitive development measures across the 1–24 month period.

In contrast to structured observations, a frequently used measure of responsivity is the Emotional and Verbal Responsivity of Mother scale from the HOME (e.g., Barnard, Bee, & Hammond, 1984; Bradley & Caldwell, 1976; Bradley & Caldwell, 1984). This scale includes items such as "mother spontaneously vocalizes to child at least twice during visit," "mother responds to child's vocalization with a verbal response," and "mother tells the child the name of some object during the visit." Responsivity is correlated with the Parental Involvement scale (r = .64 for 24 month administration), that includes items like "mother tends to keep child in visual range," and "mother 'talks' to child while doing her work." There is little distinction between these scales conceptually, operationally, and statistically. Unfortunately, evidence for specific effects has been based, in part, on single validity coefficients from tables summarizing relations between all the HOME subscales and various measures of IQ (e.g., Barnard et al., 1984; Bradley et al., 1988).

CONTINGENCY OF CONTACT

Vygotsky's theory of cognitive development has also spawned a large literature on the effects of contigency in mother–child interaction on performance of cognitive skills (e.g., Freund, 1990; Pellegrini, Perlmutter, Galda, & Brody, 1990; Rogoff, 1990) in tasks such as book reading, problem solving, and decision making. These studies support the view that contingency is an interaction process variable composed of mother, child, task, and situational characteristics. On the whole, studies examining the points of contact between a child and a developmental niche suggest that contingency is a critical feature related to IQ, and other aspects of development. Unfortunately, we know little about the contingencies present across multiple levels of a given developmental niche across a day or week.

Clearly the complexity of environmental systems per se presents a major challenge to researchers interested in their effects. Covariation within and across levels of the environment exists regardless of the measures, models, and constructs chosen, and is not eliminated by statistical control or sample selection. There is another aspect of covariation within the larger frame of developmental systems that has been largely neglected in research on environmental effects on IQ: the coactive nature of environmental and biological systems.

GENETIC LESSONS FOR ENVIRONMENTAL SPECIFICITY

A fact often neglected in environmental research is that systems views of development include not only environmental factors, but also the interplay between environmental and biological influences (e.g., Gottlieb, 1991a; Greenough & Black, 1992; Haier, Siegel, Crinella, & Buchsbaum, 1993; Sameroff, 1983). Research findings supporting systems models that incorporate biological influences (e.g., canalization, reaction norm, sensitive periods) have garnered increased attention in the developmental psychology literature (Bornstein, 1989; Bronfenbrenner & Ceci, 1993; Gottlieb, 1991b; Turkheimer & Gottesman, 1991). However, systems models that address environment–biology covariation and mutual and bidirectional influences complicate our understanding of the influences of the environment per se, regardless of which *specific* environmental parameter is under consideration (Plomin & Bergman, 1991). Therefore, in order to more fully understand the role of environment in the development of IQ, environmental models must be integrated with research on the role of genetics and other biological influences (see also Chipuer, Rovine, & Plomin, 1990 on the heritability of IQ). Therefore we consider the concept of developmental niche to include not only environmental variables, but biological influences as well. We propose two specific points of integration: research demonstrating that siblings within the same family do not share the same developmental niche, and research on environment–genetic covariation.

By highlighting the magnitude of within-family variability and its relation to developmental outcomes, behavioral genetic studies of individual differences offer an alternative model of environmental influence that deserves further study from researchers interested in cognitive development (Plomin, Chipuer, & Neiderhiser, in press). Behavioral genetic studies partition the environment into a "shared," or common, component, representing all environmental influences that serve to make siblings similar, and a "nonshared," or unique, component, representing all influences that serve to make siblings dissimilar (Plomin & Daniels, 1987). Numerous studies indicate that shared genes contribute most to sibling similarity and that siblings are dissimilar because they experience different, or "nonshared," environments. Behavior–genetic analyses of the environment suggest that distal influences (e.g., cultural beliefs and practices, SES, neighborhood, school system) are shared by siblings; however, siblings' perceptions of these factors may differ and these "shared" influences may have a different impact on siblings whose different genetic or proximal process experiences moderate similar macrosystem environments. Proximal processes such as maternal responsiveness appear more clearly to be nonshared by siblings (Dunn, Stocker, & Plomin, 1990). The shared/nonshared environment distinction is significant for highlighting the differences in developmental niches for siblings in the same family.

Distinguishing nonshared from shared environmental influence and including siblings pairs who vary in genetic relatedness (e.g., twins, adoptees) are two strategies from genetic studies that can be applied in environmental research. Environmental studies can specify the degree to which siblings share similarities in different levels of the developmental niche (e.g., maternal responsiveness vs. daycare, school). Importantly, the effects of different developmental niches can also be examined with a "genetic control," so the genetic contribution to sibling similarity can be estimated directly (e.g., by contrasting MZ twins to full siblings). Plotting the relations between sibling differences in IQ according to differences in their niche offers an alternative window on the reaction norm of IQ (Turkheimer, 1991). Admittedly, this is a rather simple design and several problems persist (e.g., gene–environment and environment–environment interactions are not addressed directly; covariance between genetic similarity and environmental similarity between siblings is not addressed). Nonetheless, employing behavior genetic strategies; particularly, studying siblings of varying degrees of consanguinity (Bronfenbrenner & Ceci, 1993), enlarges the focus of environmental studies from strictly environmental to an organized constellation of factors that shape intellectual development.

Second, a systems view that incorporates genetic and environmental influences emphasizes the correlation between genes and environment and how children influence the environments they experience (e.g., Scarr, 1992). Two aspects of gene–environment correlation are important—correlations between environment and parents' genes, and correlations between the child's genes and the environments they evoke or experience. Most often, researchers handle confounded genetic and environmental variables statistically by partialling the effects of one variable from the other, without attending to the developmental niche (environmental and biological factors). In the case of correlation between parent genes and rearing environment, maternal IQ, SES, or education level are often partialled from environmental measures (e.g. the HOME). This approach has produced inconsistent results, with the significance of environmental influences remaining in some studies but not in others (e.g., Barnard et al., 1984; Bornstein & Tamis-Lamonda, 1989; Braungart, Fulker, & Plomin, 1992; Estrada, Arsenio, Hess, & Holloway, 1987. In part, the inconsistency of results is exacerbated by the measurement problems mentioned earlier. Moreover, as previously discussed, the organization and dynamic interaction among environmental influences renders attempts to partial parental genetic influences from environmental variables difficult to interpret. Parental genes may be correlated with environmental factors that may also foster children's cognitive development dependent of genetic influence (e.g., better schools, better health care, etc.).

Evidence supporting a correlation between child characteristics and the environment children experience or evoke, particularly in terms of parental responsiveness, dates back two decades (e.g., Bell, 1968). Few researchers have examined any of the many ways in which (genetically-influenced) child characteristics, notably, temperament, influence environmental experiences that may be related to cognitive development (e.g., Lerner, 1984; Matheny, 1990; Peters-Martin & Wachs 1984). Pianta & Egeland (1994) reported that some of the instability in cognitive development on standardized tests was related to child characteristics such as frustration tolerance and symptoms of anxiety and withdrawal. These potentially genetically-related behaviors influence responses by parents (and teachers) as well as childrens' capacity to learn or benefit from environments conducive to cognitive development. However, most environment–IQ research has not included what children bring to the interactions with their parents. Bornstein (1989) and Wachs et al. (1993) correlated child's cognitive competence with maternal responsiveness to (non)distress without examining how infants who are more prone to distress may elicit more responsiveness to distress (or caregiver behavior that is not cognitively stimulating) compared to inhibited or temperamentally easy peers.

Generally, problems with covariance among and within levels of the environment are complicated by recent studies supporting gene–environment covariance. That covariance among genetic and environmental influences is the natural state of affairs is consistent with a systems conceptualization of development and further endorses the study of developmental niches rather than isolating a specific environmental or genetic influence. Moreover, it is clear that developmental changes in intelligence (McCall & Carringer, 1993), environmental (Bornstein & Tamis-LeMonda, 1989) and genetic influences (Loehlin, Horn, & Willerman, 1989), and their co-action (e.g., epigenesis) significantly complicate matters.

FUTURE DIRECTIONS AND CONCLUDING REMARKS

The hypothesis of environmental specificity offers a powerful heuristic for examining environment–IQ relations. However, we suggest that many of the studies that apply ideas of environmental specificity mis-specify the complex and dynamic nature of development. Heeding calls from Cairns (1991) for hypotheses rather than reiterating the difficulty of studying complex systems, we propose a significant shift in the level of analysis used to conceptualize and assess the environment, and integration of biologically sensitive samples and designs. More specifically, we propose research that describes developmental niches (genes *and* environment) for IQ, and hypothesize that (a) contingency assessed at the niche-level (i.e., contingency of contact between child and an array of multilevel environments) predicts cognitive development across time better than contingency at the level of specific variables, (b) specific niches have differential effects on cognitive development, and (c) predictions of IQ and developmental changes in IQ are improved by including genetically-sensitive child variables (e.g., temperament) in multi-level studies of the environment.

What would be niche-level measurement strategies? We can think of several possibilities. First, samples must be genetically-informative (e.g., twins, adoptees, siblings). Without including a *measurable* genetic factor, many of the genetic lessons and links will go untested. Second, for both conceptual and statistical reasons, studies must draw samples from a wide range of proximal environments and distal factors. Given the emphasis on moderating factors and systems of influence (Pianta & Egeland, 1994; Wachs, 1992), studies that isolate specific environmental influences will be difficult to interpret. Third, measurement strategies should depend on specific and detailed procedures and examine the covariation of environmental influences. Fourth, although procedures such as regression are helpful for isolating specific "causes", we believe that they can be inappropriate for some aspects of modeling the covarying and multilevel nature of the environment. Cluster analytic approaches and ethnographic methods provide alternative windows on the patterns of environmental and genetic influences on development and should be considered. Finally, increased attention to (a) *description* of the environment, (b) the properties of contingent contact between child and environment, and (c) the organization of environmental influences will complement ongoing research on specificity, and contribute to a more comprehensive understanding of the role of environment in the development of intelligence.

REFERENCES

Achenbach, T. M., Phares, V., Howell, C. T., Rauh, V. A., & Nurcombe, B. (1990). Seven-year outcome of the Vermont Intervention Program for Low-Birthweight Infants. *Child Development, 61*, 1672–1681.

Bakeman, R., Adamson, L. B., & Strisik, P. (1989). Lags and logs: Statistical approaches to interaction. In M. H. Bornstein and J. S. Bruner (Eds.), *Interaction in human development* (pp. 244–260). Hillsdale, NJ: Erlbaum.

Barnard, K. E., Bee, H. L., & Hammond, M. A. (1984). Home environment and cognitive development in a healthy, low-risk sample: The Seattle Study. In A. W. Gottfried (Ed.), *Home environment and early cognitive development: Longitudinal research* (pp. 117–150). New York: Academic Press.

Beckwith, L., & Cohen, S. E. (1984). Home environment and cognitive competence in preterm children during the first 5 years. In A. W. Gottfried (Ed.), *Home environment and early cognitive development: Longitudinal research* (pp. 235–272). New York: Academic Press.

Beckwith, L., & Parmelee, A. H. (1986). EEG patterns of preterm infants, home environment, and later IQ. *Child Development, 57*, 777–789.

Bell, R. Q. (1968). A reinterpretation of the direction of effects in studies of socialization. *Psychological Review, 75*, 81–95.

Bornstein, M. H. (1989). Between caretakers and their young: Two modes of interaction and their consequences for cognitive growth. In M. H. Bornstein and J. S. Bruner (Eds.), *Interaction in human development* (pp. 197–216). Hillsdale, NJ: Erlbaum.

Bornstein, M. H., & Tamis-LeMonda, C. S. (1989). Maternal responsiveness and cognitive development in children. In M. H. Bornstein, (Ed.), *Maternal responsiveness: Characteristics and consequences. New Directions in Child Development*, No. 43 (pp. 49–61). San Francisco: Jossey-Bass.

Bradley, R. H. (1989). The use of the HOME inventory in longitudinal studies of children development. In M. H. Bornstein and N. A. Krasegnor (Eds.), *Stability and continuity in mental development: Behavioral and biological perspectives* (pp. 191–216). Hillsdale, NJ: Erlbaum.

Bradley, R. H., & Caldwell, B. M. (1976). Early home environment and changes in mental test performance from 6 to 36 months. *Developmental Psychology, 12*, 93–97.

Bradley, R. H., & Caldwell, B. M. (1984). 174 children: A study of the relationship between home environment and cognitive development during the first 5 years.

In A. W. Gottfried (Ed.), *Home environment and early cognitive development: Longitudinal research.* (pp. 5–56). New York: Academic Press.

Bradley, R. H., Caldwell, B. M., & Rock, S. L. (1988). Home environment and school performance: A ten-year follow-up and examination of three models of environmental action. *Child Development, 59*, 852–867.

Braungart, J. M., Fulker, D. W., & Plomin, R. (1992). Genetic mediation of the home environment during infancy: A sibling adoption study. *Developmental Psychology, 28*, 1048–1055.

Breitmayer, B. J., & Ramey, C. T. (1986). Biological nonoptimality and quality of postnatal environment as codeterminants of intellectual development. *Child Development, 57*, 1151–1165.

Bronfenbrenner, U., & Ceci, S. J. (1993). Heredity, environment, and the question "how?"—A first approximation. In R. Plomin & G. E. McClearn (Eds.), *Nature nurture, and psychology* (pp. 313–324). Washington, DC: APA Books.

Burchinal, M., Lee, M., & Ramey, C. T. (1989). Type of day-care and preschool intellectual development in disadvantaged children. *Child Development, 60*, 128–137.

Cairns, R. B. (1991). Multiple metaphors for a singular idea. *Developmental Psychology, 27*, 23–26.

Chipuer, H. M., Rovine, M. L., & Plomin, R. (1990). LISREL Modeling: Genetic and environmental influences on IQ revisited. *Intelligence, 14*, 11–29.

Dunn, J., Stocker, C., & Plomin, R. (1990). Nonshared experiences within the family: Correlates of behavioral problems in middle childhood. *Development and Psychopathology, 2*, 113–126.

Estrada, P., Arsenio, W. F., Hess, R. D., & Holloway, S. D. (1987). Affective quality of the mother–child relationship: Longitudinal consequences for children's school-relevant cognitive functioning. *Developmental Psychology, 23*, 210–215.

Freund, L. S. (1990). Maternal regulation of children's problem solving behavior and its impact on children's performance. *Child Development, 61*, 113–126.

Gottlieb, G. (1991a). Experimental canalization of behavioral development: Theory. *Developmental Psychology, 27*, 4–13.

Gottlieb, G. (1991b). Experimental canalization of behavioral development: Results. *Developmental Psychology, 27*, 35–39.

Greenough, W. T., & Black, J. E. (1992). Induction of brain structure by experience: Substrates for cognitive development. In M. Gunnar & C. A. Nelson (Eds.), *Behavioral developmental neuroscience: Vol. 24. Minnesota symposia on child psychology*. Hillsdale, NJ: Erlbaum.

Haier, R. J., Siegel, B. V., Crinella, F. M., & Buchsbaum, M. S. (1993). Biological and psychometric intelligence: Testing an animal model in humans with positron emission tomography. In D. K. Detterman (Ed.), *Current topics in human intelligence*. Norwood, NJ: Ablex.

Lazar, I., & Darlington, R. (1982). Lasting effects of early education: A report from the consortium for longitudinal studies. *Monographs of the Society for Research in Child Development, 47*, (2/3, Serial No. 195).

Lee, V. E., Brooks-Gunn, J., Schnur, E., & Liaw, F. R. (1990). Are Head Start effects sustained? A longitudinal comparison of disadvantaged children attending Head Start, no preschool, and other preschool programs. *Child Development, 61*, 495–507.

Lerner, J. (1984). The import of temperament for psychosocial functioning: Tests of a goodness of fit model. *Merrill-Palmer Quarterly, 30*, 177–188.

Lerner, R. M. (1989). Developmental contextualism and the life-span view of person–context interaction. In M. H. Bornstein and J. S. Bruner (Eds.), *Interaction in human development* (pp. 217–243). Hillsdale, NJ: Erlbaum.

Loehlin, J. C., Horn, J. M., & Willerman, L. (1989). Modeling IQ change: Evidence from the Texas adoption project. *Child Development, 60*, 993–1004.

Marvin, R. S., & Pianta, R. C. (1992). A relationship-based approach to self-reliance in young children with motor impairments. *Infants and Young Children, 4*, 33–45.

Matheny, A. (1990). Temperament and cognition: Relations between temperament and mental test scores. In G. Kohnstamm, J. Bates, & M. Rothbart (Eds.), *Temperament in childhood.* New York: Wiley.

McCall, R. B., & Carringer, M. S. (1993). A meta-analysis of infant habituation and recognition memory performance as predictors of later IQ. *Child Development, 64*, 57–79.

Mink, I. T., Blacher, J., & Nihira, K. (1988). Taxonomy of family life styles III: Replication with families with severely mentally retarded children. *American Journal on Mental Retardation, 93*, 250–264.

Ogbu, J. U. (1987). Cultural influences on plasticity in human development. In J. J. Gallagher and C. T. Ramey (Eds.), *The malleability of children* (pp. 155–170). Baltimore: Paul H. Brookes.

Pellegrini, A. D., Perlmutter, J. C., Galda, L., & Brody, G. H. (1990). Joint reading activities between black Head Start children and their mothers. *Child Development, 57*, 443–453.

Peters-Martin, P., & Wachs, T. D. (1984). A longitudinal study of temperament and its correlates in the first 12 months. *Infant Behavior and Development, 7*, 285–298.

Pianta, R. C., & Egeland, B. (in press). Predictors of instability in children's mental test performance at 24, 48, and 96 months. *Intelligence.*

Plomin, R., & Bergeman, C. S. (1991). The nature of nurture: Genetic influences on "environmental" measures. *Behavioral and Brain Sciences, 14*, 373–427.

Plomin, R., & Daniels, D. (1987). Why are children in the same family so different from one another? *Behavioral and Brain Sciences, 10*, 1–16.

Plomin, R., Chipuer, H. M., & Neiderhiser, J. (in press). Behavior genetic evidence for the importance of nonshared environment. In E. M. Hetherington, D. Reiss, & R. Plomin (Eds.), *Nonshared environment.* Hillsdale, NJ: Erlbaum.

Ramey, C. T., Bryant, D. M., & Suarez, T. M. (1985). Preschool compensatory education and the modifiability of intelligence: A critical review. In D. Detterman (Ed.), *Current topics in intelligence* (pp. 247–296). Norwood, NJ: Ablex.

Ramey, C. T., MacPhee, D., & Yeates, K. O. (1982). Preventing developmental retardation: A General Systems Model. In L. A. Bond & J. M. Joffe (Eds.), *Facilitating infant and early childhood development* (pp. 343–401). Hanover, NH: University Press of New England.

Rogoff, B. (1990). *Apprenticeship in thinking.* New York: Oxford University Press.

Sameroff, A. J. (1989). Principles of development and psychopathology. In A. J. Sameroff & R. N. Emde (Eds.), *Relationship disturbances in early childhood* (pp. 17–32). New York: Basic Books.

Sameroff, A. J. (1983). Developmental systems: Context and evolution. In P. Mussen (Ed.), *Handbook of child psychology: Vol. 1, History, theory and methods.* (pp. 237–294). New York: Wiley.

Sameroff, A. J., Seifer, R., Baldwin, A., & Baldwin, C. (1993). Stability of intelligence from preschool to adolescence: The influence of social and family risk factors. *Child Development, 64*, 80–97.

Sameroff, A. J., Seifer, R., Barocas, R., Zax, M., & Greenspan, S. (1987). Intelligence quotient scores of 4-year-old children: Social environment risk factors. *Pediatrics, 79*, 343–350.

Scarr, S. (1992). Developmental theories for the 1990s: Development and individual differences. *Child Development, 63*, 1–19.

Super, C. M., & Harkness, S. (1986). The developmental niche: A conceptualization at the interface of child and culture. *International Journal of Behavioral Development, 9*, 545–569.

Turkheimer, E. (1991). Individual and group differences in adoption studies of IQ. *Psychological Bulletin, 110*, (3), 392–405.

Turkheimer, E., & Gottesman, I. I. (1991). Individual differences and the canalization of human behavior. *Developmental Psychology, 27*, 18–22.

Wachs, T. D. (1984). Proximal experience and early cognitive–intellectual development: The social environment. In A. W. Gottfried (Ed.), *Home environment and early cognitive development: Longitudinal research* (pp. 273–328). New York: Academic Press.

Wachs, T. D. (1992). *The nature of nurture.* Newbury Park, CA: Sage.

Wachs, T. D. (1993). Determinants of intellectual development: Single determinant research in a multidetermined universe. *Intelligence, 17*, 1–10.

Wachs, T. D., Bishry, Z., McCabe, G., Galal, O., & Shaheen, F. (1993). Relation of rearing environment to adaptive behavior of Egyptian toddlers. *Child Development, 64*, 586–604.

Wachs, T. D., & Gruen, G. E. (1982). *Early experience and human development.* New York: Plenum Press.

Wasik, B. H., Ramey, C. T., Bryant, D. M., & Sparling, J. J. (1990). A longitudinal study of two early intervention strategies: Project CARE. *Child Development, 61*, 1682–1696.

Chapter 5

Intellectual Development and the Role of Early Experience

Craig T. Ramey
Clancy Blair

Civitan International Research Center, University of Alabama at Birmingham

INTRODUCTION

Historically, intelligence has been a key concept that has provided a unifying theme for much of psychological inquiry. Early theoretical approaches tended to be concerned mainly with the structure of intellect in its mature form with Spearman (1927) emphasizing a general, or *g* factor, approach, and Thurstone (1938) and Guilford (1966) emphasizing multiple, distinct intellectual abilities.

Later, empirical efforts shifted first to identifying correlates of various tests of intelligence such as the analyses of the Fels Longitudinal Study data by McCall, Appelbaum, and Hogarty (1973), emphasizing family correlates. More recently, experimental programs, particularly for preschool-aged children, have been developed to determine the modifiability of intelligence (e.g., Lazar & Darlington, 1982; Ramey, Yeates, & Short, 1984).

The purpose of this article is to clarify our working model of intelligence and the undergirding assumptions that have guided our work on three early

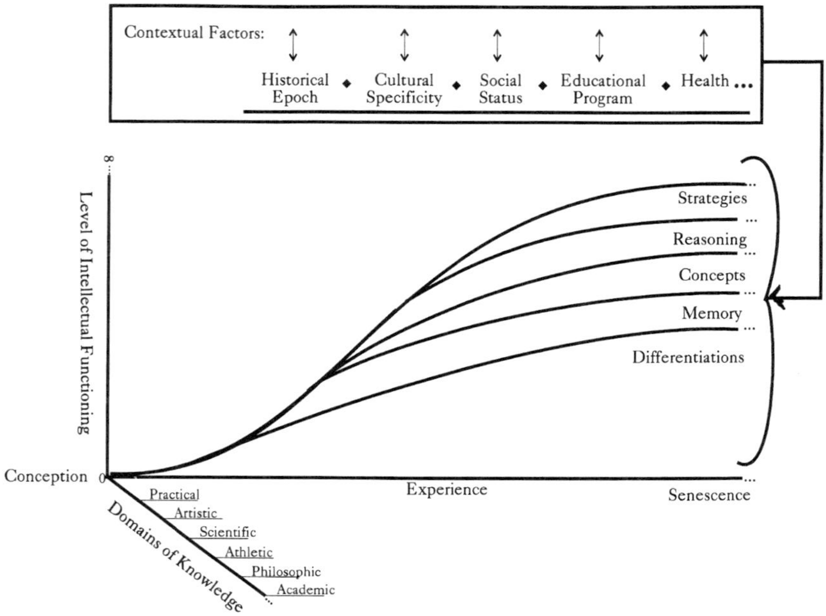

Figure 1. The Ramey–Blair inductive model of intellectual development.

intervention programs: the Abecedarian Project (Ramey & Campbell, 1992); Project CARE (Wasik, Ramey Bryant, & Sparling, 1990) and the Infant Health and Development Program (Ramey, Bryant, Wasik, Sparling, Fendt, & LaVange, 1992).

By the term intelligence, we mean the sum of component and coordinated cognitive processes that allow individuals to contemplate and achieve specific aims. These aims contain the universe of practical, artistic, scientific, athletic, philosophic, academic, and other domains of knowledge that mankind is able to produce. These domains of knowledge vary individually and culturally within a given time period and change ontogenetically for the individual, and historically for cultures and subcultures, over time.

Figure 1 contains a schematic representation of our inductive model of intelligence. The model is inductive, in that it is summative and expandable and represents our knowledge state at a given time relative to specific purposes (e.g., to develop tests to predict academic achievement).

Within our model, intelligence is construed as a developmental phenomenon, in that it changes from conception until senescence. The particular shapes of developmental curves describing the overall function of intelligence or its individual components are a matter for empirical inquiry.

We distinguish five major components of intellectual functioning which might loosely be thought of as cognitive processes. These include:

1. Differentiation—the act of distinguishing or discriminating one object, event, or element from others;
2. Memory—the act of recalling or representing a previously experienced object, event, or thought;
3. Concept Formation—the symbolic representation of a property pertaining to classes of objects, events, or symbols;
4. Reasoning—the application of a system of consistent rules (e.g., logic, mathematics, musical notation) to a particular domain of knowledge;
5. Strategic Thinking—the application of an abstract rule or system of rules that provides guidance for choices among alternative courses of action (e.g., win stay/lose shift, less is more, the calculus).

For a given domain of knowledge, intellectual growth generally takes place in the order in which the components are listed (i.e., from differentiations to strategies). Within our framework, the adequate measurement of intelligence consists of the appropriate sampling of the component cognitive processes for relevant domains of knowledge. Thus, intelligence for some individuals may be broad and evenly distributed across domains of knowledge and for others there may be varied profiles with substantial variability in accomplishment or performance among specific domains. For populations an almost infinite number of profile possibilities exist.

Domains of knowledge are interpersonally agreed upon (i.e., ultimately intersubjective) ways of thinking about large classes of related phenomenon. Domains of knowledge change over historical time (e.g., increasing knowledge about electronics following the discovery of electromagnetism) and vary from culture to culture depending on circumstances. A generally greater range of experience across domains of knowledge is likely to be associated with greater intelligence but this, too, is an empirical, rather than a philosophical, issue within our model.

Individuals vary in the rates at which they encounter various domains of knowledge and, therefore, the rate at which they acquire domain-specific differentiations, memories, concepts, reasoning and strategies. Therefore, individuals will show variation at particular ages in summative assessments of the various elements in our model which in toto we call intelligence. Within this scheme, systematic experimental variations in early experience becomes a major strategy for determining the malleability of intelligence.

Exogenous factors to our model such as genetic endowment, health, opportunity for exposure due to social class or culture, and energy level, etc., are, of course, likely to affect intellectual performance. Such factors we construe as setting conditions or contexts for the intellectual development of individuals and are represented in Figure 1 as aspects of experience that alternately constrain or expand the domains of knowledge encountered and the level of intellectual functioning attained.

The Importance of Early Experience

Early experiences within our model are important because of their logically constrained causal relationship to subsequent development. Put simply, since cause and effect are linearly related to time (and hence development), the "cause," then, for a particular developmental phenomenon must be found in the preceding experiences or contextual forces. If this premise is accepted, then the earlier one begins to study the developing child the simpler the system is and the greater the likelihood that one can understand it more clearly. A belief in the logical importance of early experience is neutral with respect to the concepts of critical or sensitive periods, which are simply descriptive terms that can be used to guide appropriately designed empirical inquiry.

The practical benefits to be gained by beginning systematic education at younger, rather than older, ages can be examined within the model of intelligence considered here. Experience, as an active agent of intellectual change, can broaden the range of domains of knowledge encountered, increase the level of intellectual functioning within a relatively narrow band of domains, or increase both breadth of domains and levels of functioning. While the universe of possible functional domains cannot be known for any one point in time, lacking omniscience, a principle of development indicates that the developing individual will become more differentiated across ontogeny and hierarchically and more flexibly organized (Werner, 1957). According to the model of intelligence under consideration here, exposure will likely produce an increase in basic cognitive skills (i.e., differentiations, memory, concepts, reasoning, and strategies) when assessed within a given domain. As development proceeds, and information within a given domain becomes increasingly complex, changes in intellectual functioning subsequent to exposure become more advanced.

Early Experience, Education, and Culture

The relationship between experience gained in a particular domain of knowledge and increases in general level of intellectual functioning has been a matter of systematic empirical enquiry. Given positive evidence for the initial effectiveness of early educational intervention programs to alter IQ for disadvantaged children (e.g., Lazar & Darlington, 1982), an important principle to be considered for program development is the selection of knowledge domains to be sampled.

The importance of this principle and the practical benefits to be gained through early educational intervention can be demonstrated with reference to intellectual development in a cross-cultural perspective. The range of experience across cultures at early ages are, we believe, more, rather than less, similar. As development proceeds however, the domains of knowledge specific to various cultures can become increasingly dissimilar. In the United States, the provision of early educational interventions has been geared to later success in public school. Preventing school failure is of major concern in the promotion of success among individuals within U.S. culture. Arguably, the promotion of academic abilities is an issue of cross-cultural concern, and perhaps, within our model of intellectual development, early educational experience might represent an effective way of promoting increases in intellectual functioning at early ages which can be maintained through continued mentoring across a variety of culture-specific domains of knowledge. But this idea requires systematic investigation.

The idea that intelligence consists of distinctions gained through experience within domains of knowledge and the application of cognitive processes across domains contains the idea that intelligence is a culturally relative phenomenon. Our model supports the notion that intelligence should be tested within culturally valid domains of functioning. Primarily, this results from the idea of cultural tutelage. That is, within a given culture, certain domains of knowledge are valued and others are considered to be of lesser value. In the range of intellectual development, this can apply not only, and perhaps not primarily, to cultural groups, but also to subgroups of a given population or culture. We suggest that intellectual development can be understood not only within the range of domains of knowledge encountered, but also with regard to experiential and motivational factors which influence the selection of knowledge domains within a given cultural subgroup. While cultural sensitivity is a worthwhile social goal, we are suggesting that even greater specificity with regard to subgroups and individuals functioning within a given subgroup is needed in conceptions of intellectual development and the developmental process.

Early Experience and Intellectual Priming Mechanisms

We assume that the young child is a naturally curious and exploratory being. We do not believe that he or she alone, however, can discover or create domain-specific knowledge until certain basic experiences have occurred. These experiences might be thought of as priming mechanisms for subsequent self-guided intellectual development. Recently, Ramey and Ramey (1992) summarized six classes of experiences that they hypothesize to be *essential* for normal cognitive development. These experiences are presented in Table 1.

Additional research needs to be conducted to determine the relationship between the frequency of occurrence of these early experiences and rate and levels of intellectual development. This research can be both correlational and experimental in form. It is correlational in that variations in patterns of occurrence of priming mechanisms in populations could be correlated with variations in the rate of development or summative level of overall intellectual func-

Table 1. Developmental priming mechanisms to promote positive cognitive development.

1. *Encouragement of exploration:* To be encouraged by adults to explore and to gather information about their environments.
2. *Mentoring in basic skills:* To be mentored (especially by trusted adults) in basic cognitive skills, such as labeling, sorting, sequencing, comparing, and noting means–ends relationships.
3. *Celebration of developmental advances:* To have their developmental accomplishments celebrated and reinforced by others, especially those with whom they spend a considerable amount of time.
4. *Guided rehearsal and extension of new skills:* To have responsible others help them in rehearsing and then elaborating upon (extending) their newly acquired skills.
5. *Protection from inappropriate disapproval, teasing, or punishment:* To avoid negative experiences associated with adults' disapproval, teasing, or punishment for those behaviors that are normative and necessary in children's trial-and-error learning about their environments (e.g., mistakes in trying out a new skill, unintended consequences of curious explorations, or information seeking). (Note: This does not mean that constructive criticism and negative consequences cannot be used for other child behaviors that children have the ability to understand are socially unacceptable.)
6. *A rich and responsive language environment:* To have adults provide a predictable and comprehensible communication environment in which language is used to convey information, provide social rewards, and encourage learning of new materials and skills. (Note: Although language spoken to the child is most important early on, the language environment may be supplemented in valuable ways by the use of written materials.)

tioning or the functioning of component processes. It is experimental in that planned variations in the six classes of experience embedded in early childhood education programs could be implemented to draw stronger causal influences about the relative importance of the various classes of experiences to *patterns* of intellectual development.

Our model of intelligence emphasizes the role of early experience in development, and malleability is understood to be a function of systematic variations in experience operating within a set of historically, biomedically, and socially defined contexts. Given our emphasis on intellectual development occurring within specified domains of knowledge, it *could* be viewed as contradictory to suggest that early educational experience for all children, particularly those from disadvantaged backgrounds, should be pursued as a national goal. An argument could be forwarded against providing enhancing experiences for disadvantaged children on the grounds that the experiences provided are limited to a necessarily restricted range of knowledge domains (e.g., academic, artistic). We emphatically do not accept this proposition but suggest it in order to make the point that developmental interventions must be ecologically valid if they are to be sustainable. While experiences gained within educational knowledge domains may improve a child's ability to succeed in school, experience might not be provided in other domains which are deemed necessary for successful intellectual functioning within a given cultural subgroup to which a child belongs. As a result, alterations in intellectual functioning associated with an intervention may not be maintainable following the offset of intervention services. Within our model social, educational, and biomedical constraints on development that the child faces in the natural ecology may conspire to limit benefits gained through intervention. This aspect of our model addresses what we believe to be a fundamental issue in early intervention research, namely, that gains associated with the provision of early educational intervention may be specific to particular domains of knowledge and consequently, that increased rates of knowledge acquisition associated with early intervention are not necessarily maintainable at later periods because of a possible ecological invalidity. We believe that one of the strengths of our model is that it offers a possible resolution of this issue through an increased emphasis on cultural specificity in intervention content.

Early Experience and the Issue of Lasting Effects

Our model provides a hypothetical basis for the expectation that early educational experience can *establish* and *maintain* an increased level of intellectual functioning. As intellectual development proceeds, domains of knowledge sampled and the level of cognitive functioning become more complex. As increasing complexity is encountered, the gains made through early intervention are hypothesized to aid the individual in negotiating new information in a variety of situations. The investigation of these hypothesized mechanisms is a matter of systematic inquiry which can inform both our knowledge of intellectual development and better inform public policy decision making concerning early intervention and its long-term outcomes.

The issue of "lasting effects" in the context of early experience must be understood as a maintenance of a *rate of acquisition* of particular domains of knowledge. For example, preschool programs may provide experiences relevant to basic mathematical concepts such as size or number, but these must be understood as prerequisites to the formal operations required for addition or multiplication and these in turn as prerequisites for mastery of algebra. Without postulation of specific "carrier mechanisms," it would seem unwarranted to expect even an excellent preschool program which had an emphasis on mathematical fundamentals to result in relative superiority, in a normative sense, in facility with algebra at, say, ninth grade. That is not to say that such continuities are impossible, but simply that there must be some bridging or *carrier mechanism* to provide the scaffolding for such a relationship. What are plausible "carrier mechanisms" for such apparent long-term continuities that have been reported in the literature (see Campbell & Ramey, 1994; Lazar & Darlington, 1982)?

We discern four major types of carrier mechanisms:

1. An addition to a child's *intellectual skills* that allows the child to control exposure of him or herself to new development-enhancing experiences, for example, through increased reading facility and accompanying access to developmentally appropriate books.
2. A *motivational* change in the child such that the child seeks out or creates advantageous learning experiences, for example, by successful competition for scholarships or their equivalent.
3. An enhanced knowledge base that results in the child's being selected for enhanced educational *tracking* (into broader or higher planes of experience) or avoidance of circumstances that punish developmental advances.
4. The child has access to more accomplished *mentors* than otherwise would have occurred (e.g., improved parenting, better teachers, more accomplished peers) who possess and use skills that provide a more optimal match between the child's developmental level and the experiences that will enhance the child's acquisition of additional intellectual breadth and development.

These carrier mechanisms warrant systematic inquiry as a high research priority.

In conclusion, intellectual development within our model is understood to result from an iterative process in which level of intellectual ability, exogenous constraints, and experiences gained in various domains of knowledge co-determine an individual's overall developmental pathway. That is, factors associated with the individual, factors associated with the environment, and experiences gained through the association of the individual and the environment result in complex psychological phenomena that are summatively labeled as intelligence. The model is iterative, in that early experience can promote developmental gains and set the stage for enhanced development, but the iterative process suggests that intelligence is developmental and, as a result, intellectual functioning can theoretically be limited or enhanced at any point during the life span. Understanding how individual and exogenous factors transact to determine development is a high priority and in this the model places great emphasis on empirical inquiry and is amenable to a broad array of structural and functional analyses. The model is consistent with the current body of early intervention research. It is inductive as it attempts to explain intellectual development in a way that is congruent with empirically observed phenomena. The primary strength of the model is that it offers a possible explanation for fading intervention effectiveness following intervention offset with reference to ecological validity and increased cultural specificity. Identifying this increased specificity and relating it to commonalities among individuals is the most obvious challenge to the model.

REFERENCES

Campbell, F. A., & Ramey, C. T. (1994). Effects of early intervention on intellectual and academic achievement: A follow-up study of children from low-income families. *Child Development, 65*, 684–698.

Guilford, J. P. (1966). Intelligence: 1965 model. *American Psychologist, 21*, 20–26.

McCall, R. B., Appelbaum, M. I., & Hogarty, P. S. (1973). Developmental changes in mental performance. *Monographs of the Society for Research in Child Development, 38*, (3, Serial No. 150), 1–84.

Lazar, I., & Darlington, R. (1982). Lasting effects of early education: A report from the Consortium of Longitudinal Studies. *Monographs of the Society for Research in Child Development, 47*, (2–3, Serial No. 195).

Ramey, C. T., Yeates, K. O., & Short, E. J. (1984). The plasticity of intellectual development: Insights from preventive intervention. *Child Development, 55*, 1913–1925.

Ramey, C. T., & Campbell, F. A. (1992). Poverty, early childhood education, and academic competence: The Abecedarian experiment. In A. Huston (Ed.), *Children in poverty* (pp. 190–221). New York: Cambridge University Press.

Ramey, C. T., Bryant, D. M., Wasik, B. H., Sparling, J. J., Fendt, K. H., & LaVange, L. M. (1992). Infant health and development program. *Pediatrics, 89*, 454–465.

Ramey, S. L. & Ramey, C. T. (1992). Early educational intervention with disadvantaged children—To what effect? *Applied and Preventive Psychology, 1*, 131–140.

Spearman, C. (1927). *The abilities of man.* New York: Macmillan.

Thurstone, L. L. (1938). *Primary mental abilities.* Chicago: University of Chicago Press.

Wasik, B. H., Ramey, C. T., Bryant, D. M., & Sparling, J. J. (1990). A longitudinal study of two early intervention strategies: Project CARE. *Child Development, 61*, 1682–1696.

Werner, H. (1957). The concept of development from a comparative and organismic point of view. In D. Harris (Ed.), *The concept of development* (pp. 000). Minneapolis: University of Minnesota Press.

Chapter 6

Environment and Intelligence: Present Status, Future Directions

Theodore D. Wachs

Department of Psychological Sciences, Purdue University

This chapter is based upon an assumption, namely that environmental influences are related, both directly and indirectly, to variability in intelligence competence. While some researchers might argue with this fundamental assumption (e.g., Scarr, 1992), for the most part it is accepted, not only by environmental researchers (Baumrind, 1993; Jackson, 1993), but by many biologically-oriented researchers, as well (e.g., Plomin, 1990; Pollitt, 1988). Within the framework of this assumption, I hope to present not only the current state of knowledge about environment and intellectual competence, but, more importantly, directions for future research in this area.[1]

[1] I am using a broad-based definition of cognitive competence, which encompasses not only IQ test performance but also adaptive behavior skills, academic performance, and performance on measures such as symbolic play which may underlie test performance.

ENVIRONMENT AND INTELLIGENCE: WHERE WE ARE

The Structure of the Environment

If we were to judge solely by the number of publications on the subject, it would be easy to conclude that the environment of the child essentially consists of direct transactions between child and caregiver (e.g., caregiver responsivity, vocalization, scaffolding). In fact, at both descriptive and process levels, the environment of the child is much more differentiated and much more complex. Current descriptive systems emphasize that the structure of the child's environment extends well beyond transactions with individual caregivers, encompassing not only noncaregiver interpersonal transactions, but also societal factors, cultural institutions and physical–ecological characteristics (Horowitz, 1987; Saco-Pollitt & Pollitt, 1994; Super & Harkness, 1986; Whiting, 1981). Of the descriptive systems available, perhaps the most well known is that of Bronfenbrenner (1989, 1993), who characterized the environment as being divided into four hierarchical but interrelated levels: the microsystem, mesosystem, exosystem and macrosystem.

It is also clear that each level of the environment is comprised of multiple subunits (Wachs, 1992). For example, the microsystem is divided into both physical and social domains (Wohlwill & Heft, 1987). The social microenvironment (characterized by transactions between caregiver and child) can be further divided into a series of subdomains that are definable on the basis of source, modality, and characteristics of caregiver input (Bradley & Caldwell, in press). Similarly, the physical microenvironment (the stage or setting upon which caregiver child transactions take place) can also be divided into various subdomains, defined along such dimensions as the degree of stimulus salience and responsivity (Wachs, 1989), affordances offered by stimuli (Gibson, 1979) and spatial characteristics (Moore, 1987).

Conceptually, the different levels and subunits of the environment are often treated as distinct. In reality, these levels and subunits are bidirectionally interconnected, both within and across levels (Wachs, 1992). These interconnections mean that environmental characteristics at one level or subunit may moderate the influence of environmental characteristics at a different level or subunit; they also mean that environmental characteristics at one level or subunit may change the nature of environmental characteristics at a different level or subunit. As one example of cross-level moderation, relations between adolescent school achievement and authoritative parental rearing styles (microsystem) vary as a function of both macrosystem based ethnic–cultural factors and mesosystem-based peer group values (Steinberg, Dornbush, & Brown, 1992). Similarly, relations between children's cognitive performance and the exosystem factor of parental occupational level is moderated by higher-order macrosystem factors such as residence area (Goduka, Poole, & Aotaki-

Phenice, 1992). As one example of within-level influences, higher levels of noise-crowding in the home have repeatedly been shown to be associated with lower levels of caregiver vocalization, responsivity, and scaffolding of the environment for the child (Wachs, 1993, Wachs & Camli, 1991).

What the above clearly shows is that our understanding of the nature of the child's environment has grown well beyond simple models of caregiver child transactions. Based on evidence illustrating the existence and importance of multilevel interrelated environmental domains I have argued that it may be more appropriate to talk about *environmental systems*, rather than *the environment*, when discussing relations between environment and development (Wachs, 1992).

The Effective Environment and Cognitive Competence

Besides an increased understanding of the nature of the environment, available research has also documented linkages between specific dimensions of the environmental system and various domains of cognitive competence. Among the dimensions of the environmental system that have been reliably associated with variability in cognitive competence are: (a) cultural beliefs about what cognitive components are important for adaptation; (b) school attendance; (c) school or parental expectancy for achievement; (d) availability, variety, and responsivity of social and object stimulation; (e) parental involvement; (f) caregiver-guided participation of the child in problem-solving activities; and (g) lower levels of noise, crowding, or restriction of exploration (for reviews of this area see Belmont, 1989; Ceci, 1991; Rogoff, 1990; Wachs, 1992; Wachs & Gruen, 1982).

While many of the dimensions highlighted above have been identified primarily through correlational research, a causal role for environment in the development of cognitive competence is established by reference to two other areas of research. First, there are experimental intervention studies (e.g., Achenbach, Howell, Aoki, & Rauh, 1993; Brooks-Gunn, Klebanov, Liaw, & Spiker, 1993; Garber, Hodge, Rynders, Dever, & Velu, 1991; Grantham-McGregor, Schofeld, & Powell, 1987; Ramey & Campbell, 1987), as well as controlled studies of adopted children (e.g., Dumaret, 1985), both of which show gains in cognitive performance associated with provision of increased or different types of environmental stimulation. Secondly, there are infrahuman studies, clearly demonstrating that variability in environmental stimulation is associated with variability in the development and functioning of central nervous system processes underlying cognitive competence (Diamond, 1988; Greenough & Black, 1992). Taken together, these multiple research lines not only indicate that environment is relevant to variability in cognitive development, but also which aspects of the environmental system appear to be most salient for cognitive competence.

Process Aspects of the Environmental System

By process aspects I refer to our understanding of how the environment works: the processes that govern how variability in environment is translated into variability in cognitive competence. At least four process variables have been documented up to the present time (for a more detailed review of this area see Wachs, 1992).

First, it is important to understand that the environmental system does not operate in isolation from biosocial factors that also can influence cognitive competence. Environmental characteristics *covary* with a variety of other potential developmental determinants including, but not limited to, individual *genetic characteristics* (Plomin, DeFries, & Loehlin, 1977), *nutritional status* (Engle, 1990; Pollitt, 1988), *morbidity* (Joffe, 1982), and exposure to *environmental toxins* (Scarr, 1985). Covariance processes may be *passive* in nature—certain combinations of environmental and biological characteristics naturally go together, as shown in evidence indicating that poorly nourished children are much more likely to be raised in inadequate rearing environments (Engle, 1990; Sigman, Neumann, Jansen, & Bwibo, 1989). Alternatively, covariance may also be *reactive*, such that children with certain biological characteristics have a higher probability of eliciting certain types of reactions from their caregivers (e.g., more adequately nourished children are more likely to elicit active involvement from their caregivers, Chavez & Martinez, 1982).[2]

One major implication of the existence of covariance between environmental and biological parameters is the need to remember that "environmental influences" upon cognitive competence must always be interpreted within the context of *covarying biological parameters*, which may also be relevant for development. The alternative implication also holds, of course; when considering "biological influences" on development it is important to recognize that their contributions must always be interpreted within the context of covarying environmental parameters.

Secondly, in addition to covariance, it is also important to recognize that environmental influences can *interact* with individual characteristics to influence the course of cognitive development. The term interaction refers to individual differences in reactivity to objectively similar environmental situations (see the chapters contained in Wachs & Plomin, 1991, for an extended discussion of the characteristics and measurement of organism–environment interaction). Interactions among predictors can be *multiplicative* (nonlinear) in nature—the combined impact of a set of predictors is greater than the summed

[2] There is a third type of covariance, namely active covariance, wherein children possessing certain characteristics have a higher probability of selecting certain types of environmental niches for themselves. While an exciting concept, at present evidence for the actual operation of active covariance is relatively scarce, at least at the human level (for a recent example illustrating active covariance see Gunnar, 1994).

impact of the individual predictors (synergistic interactions). Alternatively, protective dimensions of the environment may block the impact of biological or environmental risk factors (*buffering* interactions).

A number of examples of these types of interactive processes can be found in the cognitive domain. Significant increments in the prediction of cognitive performance have been found to be associated with nonlinear, multiplicative interactions between quality of caregiving and children's initial developmental status (Ho, 1987) or children's nutritional intake (Wachs, Moussa, et al., 1993). More adequate caregiving environments have been found to buffer (protect) the subsequent cognitive performance of children who are at risk, due either to birth complications (Breitmeyer & Ramey, 1986), or deviant electroencephalogram patterns (Beckwith & Parmalee, 1987); similarly, adequate nutrition has been shown to buffer the cognitive performance of children at risk, due either to less adequate caregiving environments (Wachs, Moussa, et al., 1993), or to general psychosocial disadvantage (Pollitt, Gorman, Engle, Martorell, & Rivera, 1993). Interactions may also occur as a function of time (age), as seen in concepts such as sensitive periods (Bornstein, 1989) and "canalization" (McCall, 1981). Temporal interactions occur when there are periods of development where the organism is more or less sensitive to environmental influences then at other periods. While temporal interactions are typically thought of as biologically driven, Gottlieb (1991) has presented convincing evidence that canalization processes also may be governed by environmental influences. The existence of these types of interactions calls into question main effect models that emphasize just environmental or just nonenvironmental contributions to development.

A *third* identified process is *specificity*. Specificity refers to different aspects of the environment relating to different aspects of development (see Wachs, 1992 for an extended discussion of specificity). As an example of specificity in the cognitive domain, data by Jennings and Conners (1989) indicated that children's verbal ability was predicted primarily by maternal affective tone, whereas children's nonverbal ability was related primarily to maternal directiveness; affective tone was unrelated to nonverbal ability and directiveness was unrelated to verbal ability. Of particular interest is recent evidence indicating that the processes governing specificity of environmental action occur not only in Western cultures, but also in non-Western cultures such as Japan (Bornstein, Azuma, Tamis, Le Monda & Ogino, 1990) and Egypt (Wachs, Bishry et al., 1993).

The concept of specificity helps to explain what some researchers regard as a major gap in current environmental theories—namely, why sibs reared together can turn out so differently. It has been argued that explanation of this phenomenon requires a new type of environmental process, namely "*nonshared*" environmental action (Plomin & Daniels, 1987). Nonshared environment refers to those environmental domains that act to make individuals different rather than similar. However, if sibs within the same family are encountering distinctive environments (see Hoffman, 1991), do we need to postulate a new environmental process? If different environments are associated with different patterns of developmental outcome (environmental specificity) we would expect a lack of developmental similarities for sibs reared under the same roof if these sibs encounter different environments.

Finally, the operation of covariance, interaction and specificity leads to the conclusion that relations between environment and cognition must be viewed within a *probabilistic framework*. (This conclusion holds not just for cognitive competence, but for most other areas of development as well: Gottlieb, 1991; Oyama, 1985). The fact that environment relates in systematic ways to development does not necessarily mean that the same environmental variables are related to cognitive competence across all contexts, or for all individuals, or for all cognitive outcomes. Rather, the relation to cognitive competence of specific environmental parameters is always likely to be moderated by higher-order contextual characteristics, individual biological characteristics (which may either covary or interact with the environment), and the specific cognitive outcomes that are under study.

One implication of the probabilistic nature of relations between environment and development is that theories about the nature of environmental influences must be modified to take into account existing complexities. A second implication is that the nature of our research activities should mirror the multidetermined nature of the environmental system, namely collaborative research between scientists from both the biological and environmental domains. Unfortunately, while recent theories of environmental action have gone beyond simple-minded parsimony to encompass multiple levels of the environment and multiple determinants of development (e.g., Bronfenbrenner, 1989, 1993; Horowitz, 1987; Wachs, 1992), the same can not be said for existing research practice. There still appears to be a clear gap between theory and practice, in the sense that current theories are becoming more multideterministic, whereas most researchers continue to work in isolation from researchers in other potentially relevant domains (Wachs, 1993b).

FUTURE DIRECTIONS IN RESEARCH INTO THE ENVIRONMENTAL SYSTEM: A PEEK INTO THE CRYSTAL BALL

Some future questions can best be addressed directly by environmentally oriented researchers. However, there are other research questions that can only be addressed by active collaboration between biologically and environmentally oriented researchers. While prophets rarely profit (certainly not within their own country), the remainder of the present chapter will be devoted to my prophecies on the most fruitful directions for future research on environment and cognition.

For Environmentalists Only: The Nature of Context

Even we though we have gone beyond simple main effect theories of environmental action to more system-based approaches, for the most part current theories (*including my own*) are primarily *descriptive* rather than dynamic in nature. Current theories do identify potentially relevant contextual and microenvironmental processes; however, they provide little guidance in terms of the processes underlying interrelations among the different levels of the environment, or how the interrelating processes are translated into individual variability in outcome. (One notable exception is a theoretical approach recently developed by Bronfenbrenner & Ceci, in press.) An excellent illustration of how our current theories are descriptive rather than dynamic in nature is provided by the question of when higher order contextual influences do or do not moderate the impact of microenvironmental proximal processes.

As noted earlier, the hypothesis that contextual factors will moderate the impact of microenvironmental processes is central to many of the recent theoretical approaches to conceptualizing environment and the role of environmental influences. A number of studies support this central hypothesis (for a more detailed review see Bronfenbrenner, 1993 or Wachs, 1992). For example, Bradley et al. (1989) have reported that the pattern of relations between caregiver behavior and children's cognitive performance varied across ethnic groups, even after statistically controlling for socioeconomic status. Stevenson and Lee (1990) have argued that cultural values stressing effort and education may buffer the academic performance of Asian children against the negative influences of overcrowded classrooms. The existence of contextual moderation is not surprising, given evidence indicating that different macrosystem or contextual settings can vary across multiple dimensions including (but not limited to): (a) differences in *parental belief systems* about the importance for development of verbal responsiveness to infants (Richman, Miller, & LeVine, 1992), or parental reinforcement of children's behavioral competancies (LeVine, 1980); (b) differences in *parental behaviors* thought to be relevant for development, such as the degree of face-to-face interaction between infants and their primary caregivers (Whiting & Edwards, 1988) or joint infant–adult verbal and nonverbal interaction with objects (Bakeman, Adamson, Konner, & Barr, 1990); (c) differences in the nature and quality of *rearing conditions*, as in the amount of caregiving of younger children by older sibs (Joekes, 1989; Rogoff, Mistry, Goncu, & Mosier, 1991).[3]

---

[3] Cultures also differ in the abilities they use to define "intelligence." Particularly in non-western cultures, definitions of intelligence often include a strong social component (Nsamenang, 1992). Based on the specificity principle discussed earlier, we might expect environmental components that are related to social outcomes in Western countries to be related to socially loaded cognitive outcomes in non-Western cultures.

Unfortunately, at least for environmental theorists, there are also exceptions to this central hypothesis. Looking just at cognitive competence, similar patterns of relations between parental object or social stimulation and infant object or social exploration have been shown in the United States, France, and Japan (Bornstein et al., 1990, Bornstein, Tamis-LeMonda, Pecheax, & Rahn, 1991). Similar patterns of relations between toddler adaptive behavior and caregiver vocal stimulation, physical contact stimulation, and nonverbal responsivity to toddler vocalization have been shown in the United States, Kenya, and Egypt (Wachs, Bishry, et al., 1993). Data from Mexico on the relation between cognitive performance and verbal responsivity or physical contact stimulation basically parallels findings from the United States on these dimensions (Miller, LeVine, & Richman, 1992). Measures of crowding, physical stimulation offered by the environment and encouragement of infant exploration, have been shown to be related in similar ways to later school achievement in both the United States and Israel (Ninio, 1990). Parental interference has been shown to be negatively related to both Pakistani and Scottish children's performance on seriation tasks (Adjai, 1977). Degree of flexibility of family structuring has been shown to be curvilinearly related to children's cognitive performance in Brazil, France, the Ivory coast, Pakistan, and Scotland (Oglivy, 1990).

The above exceptions do not obviate the fact that there are many situations wherein contextual processes do moderate the impact of microsystem factors. What these exceptions do indicate is that it critical for environmental researchers to understand the specific conditions that make it possible to find similarities or differences in microenvironment–development relations across different contexts. Understanding when contextual moderation will or will not occur would be a major step in terms of moving from descriptive to dynamic environmental systems theories.

Providing the necessary information will require a greater emphasis on studying patterns of environment–development relations in different contextual settings. This does not necessarily mean doing crosscultural research. Different contextual settings also can be found within a given culture, as in comparisons among different ethnic groups or between individuals differing on level of education or sociodemographic status (e.g., Bradley et al., 1989; Dornbush, Ritter, Leiderman, Roberts, & Fraleigh, 1987). Whether we look within or across cultures, what is needed are studies that offer not only detailed descriptions of the child's microenvironment, but also detailed descriptions of the nature of higher-order contextual processes that may directly or indirectly impinge on children's development. Studies that offer detailed analyses of the nature and impact of the child's microenvironment in a given context will be useful for understanding contextual moderation when they also provide us with evidence illustrating how and when contextually driven stresses, support systems, or caregiver beliefs enhance or attenuate the impact of specific microenvironmental processes (e.g., Richter's 1989 description of the interrelation of family structure, caregiver beliefs, and segregated housing practices as these impact upon the relation of crowding to children's cognitive development in South Africa). To do this it will be necessary to go beyond contextual social addresses to provide *detailed descriptions* on the nature of higher order contextual factors, akin to the detailed descriptions we provide of microenvironmental factors. Using exposure to schooling as an example, it is essential that researchers go beyond the label of school and provide information about the cost of schooling for the family, teacher training, and the physical set up of the classroom, as well as nonschool pressures that can impact on the teachers functioning in the classroom, such as the distance the teacher has to travel to get to school (see Gorman & Pollitt, 1992, for an example of this type of analysis). Similarly, studies focused on the role of contextual processes also need to identify the *specific proximal microenvironmental characteristics* that relate to variability in cognitive development within that context (e.g., Stevenson's 1982 study on the relation to cognitive performance of parental education, parental literacy, child schooling, and quality of the home environment in three subcultural groups).

In doing this type of contextual research, two points are important to consider. *First*, while many environment–development studies have been done using individuals from different ethnic groups or sociodemographic backgrounds, the traditional practice in these studies is to *statistically partial out* from the analyses differences in sociodemographic status, parental educational level or ethnicity. Rather than illuminating the nature of relations between higher order contextual and microenvironmental process, this strategy actually serves to conceal existing linkages (Bronfenbrenner, 1993). Rather than partialling, separate analyses relating microenvironment to development should be conducted within each context, as a means of highlighting possible contextual moderators. Obviously this is a strategy that may result in lower power, due to smaller sample sizes within subgroups (e.g., Steinberg, Mount, Lanborn, & Dornbusch, 1991). Strategies do exist for dealing with this power problem, including the use of aggregated data, the use of environmental marker variables in different studies allowing us to combine results from multiple studies, and the use of statistics that rely on confidence intervals rather than on a single true–false decision (see Wachs & Plomin, 1991 for detailed discussion of these issues). The critical point is that fear of low power per se should not stop environmental researchers from doing crosscontextual research.

Second, in carrying out these types of research strategies, environmental researchers should not feel ashamed if their research has an inductive rather than a deductive flavor. Current environmental systems theories are useful in describing the types of contextual settings we need to be considering along with our more traditional microenvironmental variables. However, it is very difficult to use existing theories to develop theory driven research about when contextual moderation may or may not occur. What is now needed are exploratory studies, designed to identify the specific conditions under which contextual moderation does or does not occur. The results of these inductive studies will form the basis for the next leap forward in environmental systems theories that will allow us to derive and test precise predictions about what types of interactions between contextual and microenvironmental processes influence cognitive development.

Going Beyond the Environment: Context Can Also Be Biological

Just as researchers investigating microenvironmental processes need to broaden their horizon to systematically integrate contextual factors into their research, so environmental researchers in general need to broaden their horizon to systematically integrate biological parameters into their research designs. The rationale for this statement is based on the points raised previously, illustrating covariance and interaction between biological and psychosocial environmental processes. We need to look at *biological contexts*, as well as environmental contexts, in order to fully understand the nature of environmental action. The emphasis in the next section is on biological domains that offer the best opportunity for collaborative biological–environmental research.

The New Genetics and Environmental Research

Traditionally, genetics is the biological parameter that is most often considered by environmentally oriented researchers. While all too much energy has been squandered on the perennial and useless nature versus nurture dichotomy, the concept of gene–environment interaction offers a real hope of both rapprochement and progress in the joint study of behavioral genetics and environment (Wachs, 1993 b). While the concept of gene–environment interaction is familiar to environmental researchers, the nature of the gene portion of this equation is being changed radically by new genetic technology. Up to the present, most studies of "gene"–environment interaction at the human level have been somewhat of a misnomer. For the most part, the gene in this equation refers to behavioral (e.g., temperament) or demographic (e.g., parental psychopathology) *phenotypes*, which are considered to have a strong genetic basis. While much of the research in this area involves noncognitive outcomes, examples of this type of design are also found in the cognitive domain, as in studies showing differential reactivity to the environment as a function of biological parent IQ (Willerman, 1979) or child temperament (Wachs & Gandour, 1983). Going beyond gene-related phenotypes, recent advances in molecular genetics may soon allow researchers to identify specific combinations of *genetic markers* associated with particular behavioral traits (see Plomin, 1990, or Plomin & Saudino, 1994, for reviews of this research).

With the new technology of molecular genetics, it may be possible to classify individuals based on their genotypes, thus allowing much more precise studies of gene–environment interaction. Given these advances, the problem in doing collaborative research on gene–environment interaction will lie not so much in the methodology as in the motivation. As I have noted elsewhere (Wachs, 1993 b) there has been disappointingly little *collaborative research* between behavior geneticists and environmentally oriented researchers. Finding ways to initiate collaborative research on gene–environment interaction is essential if we are to be able to answer such questions as the nature and extent of environmental buffering for individuals at genetic risk for developmental disorders (Rutter, 1983), or under what contextual conditions can genetic influences be maximized or minimized (Bronfenbrenner & Ceci, in press).

Environment and Nutrition

While individual genotypes form an essential portion of the individual's biological context, genotype is not the only portion. Chronic undernutrition and micronutrient deficiencies are a major aspect of the biological context of most children in less developed countries (Pollitt, 1988; Simeon & Grantham-McGregor, 1990), as well as disadvantaged children in the United States (Karp, 1993). While both nutrition and environment contribute to variability in cognitive competence, environmental researchers have shown very little interest in nutritional contributions. This neglect is particularly surprising given the strong covariance between inadequate nutrition and inadequate rearing environments, as documented earlier in this chapter.

Greater emphasis by environmentally oriented researchers on understanding the nutritional context of the child is essential for several reasons. *First*, if there is covariance between rearing environment and nutritional status, it becomes difficult to understand the role of the environment without understanding the role of covariates such as nutritional status (and vice versa). This is particularly true if the covariance is of the reactive kind, with more adequately nourished children being more likely to elicit developmentally facilitative transactions with their caregivers (Pollitt et al., 1993). As noted earlier, statistically covarying out "nuisance" variation associated with nutrition is not an appropriate strategy.

Secondly, if there are interactions between nutrition and environment, focusing just upon environmental parameters may well lead to misleading conclusions about the nature of the processes involved in environment–cognition relations. While rare, cognitive outcome studies demonstrating such interactions do exist. At the level of the microenvironment, both buffering and synergistic interactions between nutritional status and caregiver interactions patterns have been demonstrated (Wachs, Bishry, et al., 1993). In terms of higher-order contextual factors, a recent monograph by Pollitt, et al., (1993) dramatically illustrates how long term differences in cognitive functioning are associated both with buffering and synergistic interactions between nutritional interventions, sociodemographic status, and duration of educational experience. *Finally*, from an applied perspective, cognitive gains are more likely to result when interventions with malnourished children include both nutritional and psychosocial intervention strategies (Grantham-McGregor, 1993; Super et al., 1981). Given the various possible ways in which environment and nutritional status interrelate, collaborative research between nutritional and environmental researchers would seem a highly fruitful future direction, with potential benefits not only at the scientific level, but also in terms of improving children's functional competence.

Environment and Neuroscience

Within a systems framework, relations between developmental predictors at different levels should be bidirectional (Gottlieb, 1991). The previous two sections have illustrated how the results of environmental research should be interpreted within the framework of the individual's biological context. In the final section I suggest how the results of biological research can benefit from consideration of the environmental context.

Conceptual linkages between environmental research and the rapidly growing field of neuroscience come out of infrahuman studies on the role of the environment on the development and functioning of the central nervous system (Diamond, 1988; Greenough & Black, 1992). I refer specifically to the concept of *experience-dependent development.* As defined by Greenough and Black (1992), experience-dependent development encompasses neural processes involved in the storage of information that is unique to the individual. Experience-dependent development is characterized at the neural level by the production and strengthening of new synaptic connections. Research with infrahuman populations also has demonstrated that experience-dependent developmental processes can occur across the lifespan in response to unique information. Experience-dependent development is contrasted with experience-expectant development. Experience-expectant developmental processes involve selective retention of existing neural synapses as a function of exposure to stimulation that is encountered by most members of a species. Unlike experience-dependent development, experience-expectant developmental processes typically occur within a relatively limited time span.[4] While experience-dependent devel-

---

[4] The distinction between experience-dependent and experience-expectant developmental processes is also important in understanding a recent salvo fired in the recurring nature versus nurture conflict. I refer specifically to Scarr's (1992) contention that nonextreme environmental variation is essentially irrelevant to development. If we are looking only at developmental processes governed by experience-expectant development, then Scarr's conclusion is probably correct, in the sense that humans as a species are adapted to expect experience-expectant stimuli to occur at certain developmental periods, and these stimuli have a high probability of occurring during these developmental periods. However, when we look at developmental processes governed by experience-dependent development, then Scarr's contention would be incorrect, because environmental stimulation unique to each individual is responsible for this type of development.

opment has been studied primarily with infrahuman species, it would be surprising if such processes do not occur at the human level as well. If experience-dependent development can be shown to operate at the human level, there are major opportunities for collaboration between environmental and neuroscience researchers, particularly if the technology of neuroscience advances to the point that we can actually measure the operation of experience-dependent biological processes, such as formation of specific synapses or the speed of transmission across specific synaptic connections.

Given that experience-dependent development is directly related to the characteristics of unique information encountered by the individual, there are a number of critical questions that would need to be answered to illustrate the nature of these processes at the human level. Certainly, it would be important to know what types of experiences are associated with the production or functioning of specific synaptic connections. Along the same lines, it would also be important to understand whether certain types of environmental circumstances may inhibit the production or functioning of experience-dependent developmental processes (e.g., environmental chaos, confusion, noise). A second profitable area of research would derive from earlier studies, showing stabilization of early learning processes through repeated exposure (reinstatements) of previously learned cues (Rovee-Collier, 1984). If experience-dependent developmental processes are sensitive to environmental input, then it is important to know what part repeated experiences (reinstatements) play in this process, both in the development and maintenance of specific synaptic connections. Ultimately, it may be important to also consider the question of whether experience-dependent neural processes are directly linked to environmental input, or are moderated via individual characteristics of the organism (e.g., organism–environment interaction, but at a neural, rather than a behavioral, level).

CONCLUSIONS

What I have attempted to do in this chapter is provide an overview of relations between environment and cognitive competence, both in terms of the current status of the field and in terms of potential future directions. The relevance of environmental factors to cognitive development has been a fruitful area of study in the past and will continue to be so, particularly if we can develop collaborative relations with researchers from other domains. In terms of future directions, especially, the key word appears to be *context*. As used here, context refers not only to the larger environmental context within which the individual functions, but also to the biological context that is unique to each individual. Viewing the study of environment and cognitive development within a expanded contextual framework emphasizes not only the importance of increasing our knowledge about the complex nature and interplay of environmental forces that impinge on the individual, but also the necessity for developing collaborative relations with colleagues in other domains that form the biological context of the individual.

REFERENCES

Achenbach, T., Howell, C., Aoki, M., & Rauh, V. (1993). Nine year outcome of Vermont intervention program for low birth weight infants. *Pediatrics, 91*, 45–55.

Adjai, K. (1977). Influences of specific maternal occupation and behavior on Piagetian cognitive development. In P. Dasen (Ed.), *Piagetian psychology: Cross cultural contribution*, (pp. 227–256). New York: Gardner Press.

Bakeman, R., Adamson, L., Konner, M., & Barr, R. (1990). Kung infancy: The social context of object exploration. *Child Development, 61*, 794–80.

Baumrind, D. (1993). The average expectable environment is not good enough. *Child Development, 64*, 1299–1317.

Beckwith, L., & Parmelee, A. (1986). EEG patterns of preterm infants, home environments and later IQ. *Child Development, 57*, 777–789.

Belmont, J. (1989). Cognitive strategies and strategic learning. *American Psychologist, 44*, 142–148.

Bornstein, M. (1989). Sensitive periods of development *Psychological Bulletin, 105*, 179–187.

Bornstein, M., Azuma, H., Tamis-LeMonda, C., & Ogino, M. (1990). Mother and infant activity and interaction in Japan and in the United States. *International Journal of Behavioral Development, 13*, 267–288.

Bornstein, M., Tamis-LeMonda, C., Pecheux, M., & Rahn, C. (1991). Mother and infant activity and interaction in France and the United States. *International Journal of Behavioral Development, 14*, 21–44.

Bradley, R., & Caldwell, B. (in press). *Caregiving and the regulation of child growth and development: Describing proximal aspects of caregiving systems*. Developmental Review.

Bradley, R., Caldwell, B., Rock, S., Ramey, C., Barnard, K., Gray, C., Hammond, M., Mitchell, S., Gottfried, A., Seigel, L., & Johnson, D. (1989). Home environment and cognitive development in the first three years of life: A collaborative study involving six sites and three ethnic groups in North America. *Developmental Psychology, 25*, 217–235.

Breitmeyer, B., & Ramey, C. (1986). Biological nonoptimality and quality of postnatal environment as codeterminants of intellectual development. *Child Development, 57*, 1151–1165.

Bronfenbrenner, U. (1989). Ecological systems theory. *Annals of Child Development, 6*, 187–249.

Bronfenbrenner, U. (1993). Ecological systems theory. In R. Wozniak & K. Fisher (Eds.), *Specific environments: Thinking in context* (pp. 3–46). Hillsdale, NJ: Erlbaum.

Bronfenbrenner, U., & Ceci, S. (in press). *Nature–nurture reconceptualized in developmental perspective*: A bio-ecological model. Psychological Review.

Brooks-Gunn, J., Klebanov, P., Laiw, F., & Spiker, D. (1993). Enhancing the development of low birth weight premature infants: Changes in cognition and behavior over the first three years. *Child Development, 64*, 736–753.

Ceci, S. (1991). How much does schooling influence general intelligence and its cognitive components. *Developmental Psychology, 27*, 703–722.

Chavez, A., & Martinez, C. (1982). *Growing up in a developing community*. Guatemala City: Institute of Nutrition in Central America and Panama.

Diamond, M. (1988). *Enriching heredity: The impact of the environment on the anatomy of the brain*. New York: Free Press.

Dornbush, S., Ritter, P., Leiderman, P., Roberts, D., & Fraleigh, M. (1987). The relation of parenting style to adolescent school performance. *Child Development, 58*, 1244–1257.

Dumaret, A. (1985). IQ scholastic performance and behavior of sibs raised in contrasting environments. *Journal of Child Psychology and Psychiatry, 26*, 553–580.

Engle, P. (1990). *Child caregiving and infant preschool nutrition*. Paper prepared for the Cornell Food and Nutrition Policy Project and the Rockefeller Foundation. Polytechnic State University, San Luis Obispo, California.

Garber, H., Hodge, J., Rynders, J., Dever, R., & Velu, R. (1991). The Milwaukee Project: Setting the record straight. *American Journal of Mental Retardation, 95*, 493–525.

Gibson, J. (1979). *The ecological approach to visual perception*. Boston: Houghton Mifflin.

Goduka, I., Poole, D., & Aotaki-Phenice, (1992). A comparative study of Black South African children from three different contexts. *Child Development, 63*, 509–525.

Gorman, K., & Pollitt, E. (1992). School efficiency in rural Guatemala. *International Review of Education, 38*, 519–524.

Gottlieb, G. (1991). Experiential canalization of behavioral development. *Behavioral Psychology, 27*, 4–13.

Grantham-McGregor, S. (1993). Assessment of the effects of nutrition on mental development and behavior in Jamaican studies. *American Journal of Clinical Nutrition Supplements, 57*, 303s–309s.

Grantham-McGregor, S., Schofield, W., & Powell, C. (1987). Development of severely malnourished children who received psychosocial stimulation. *Pediatrics, 79*, 247–254.

Greenough, W., & Black, J. (1992). Induction of brain structure by experience: Substrates for cognitive development. In M. Gunnar & C. Nelson (Eds.), *Developmental behavior neuroscience* (pp. 153–23). Hillsdale, NJ: Erlbaum.

Gunnar, M. (1994). Psychoendocrine studies of temperament and stress in early childhood. In: J. E. Bates, T. D. Wachs (Eds), *Temperament: Individual differences at the interface of biology and behavior* (pp 175–198). Washington DC: APA Books.

Ho, H. (1987). Interaction of early caregiving environment and infant developmental status in predicting subsequent cognitive performance. *British Journal of Developmental Psychology, 5*, 183–191.

Hoffman, L. (1991). The influence of the family environment on personality. *Psychological Bulletin, 110*, 187–203.

Horowitz, F. (1987). *Exploring developmental theories*. Hillsdale NJ: Erlbaum.

Jackson, J. (1993). Human behavior genetics. Scarr's theory and her views on intervention. *Child Development, 64*, 1318–1332.

Jennings, K., & Conners, R. (1989). Mothers Interactional style and children's competence at three years. *International Journal of Behavioral Development, 12*, 155–175.

Joekes, S. (1989). Women's work and social support for child care in the third world. In J. Leslie & M. Paolisso (Eds.), *Womens work and child welfare in the third world* (pp. 59–86). Boulder: WestView Press.

Joffe, J. (1982). Approaches to prevention of adverse developmental consequences of genetic and prenatal factors. In L. Bond & J. Joffe (Eds.), *Facilitating infant and early childhood development* (pp. 121–158). Hanover: University Press of New England.

Karp, R. (1993). *Malnourished children in the U.S.* New York: Springer.

LeVine, R. (1980). Anthropology and child development. In C. Super & S. Harkness (Eds.), *Anthropological perspectives on child development* (pp. 71–86). San Francisco: Jossey-Bass.

McCall, R. (1981). Nature–nurture and the two realms of development. *Child Development, 52*, 1–12.

Miller, P., LeVine, R., & Richman, A. (1992, May). *Maternal responsiveness to infants and subsequent cognitive development: A cross cultural study.* Paper presented to the International Society for Infant Studies. Miami, Florida.

Moore, G. (1987). The physical environment and cognitive development in child care centers. In C. Weinstein & T. David (Eds.), *Spaces for children* (pp. 41–72). New York: Plenum.

Ninio, A. (1990). Early experience and school achievement in the second grade: An Israeli study. *International Journal of Behavioral Development, 13*, 1–22.

Nsamenang, A. (1992). *Human development in cultural context.* Newbury Park: Sage.

Oglivy, C. (1990). Family type and children's cognition in two ethnic groups. *Journal of Cross Cultural Psychology, 21*, 319–334.

Oyama, S. (1985). *The ontogeny of information.* Cambridge: Cambridge University Press.

Plomin, R. (1990). The role of inheritance in behavior. *Science, 248*, 183–188.

Plomin, R., & Daniels, D. (1987). Why are children in the same family so different from each other? *Behavior and Brain Sciences, 10*, 1–16.

Plomin, R., DeFries, J., & Loehlin, J. (1977). Genotype environment interaction and correlation in the analysis of human development. *Psychological Bulletin, 84*, 309–322.

Plomin, R., & Saudino, K. (1994). Quantitative genetics and molecular genetics. In J. E. Bates & T. D. Wachs (Eds.), *Temperament: Individual differences at the interface of biology and behavior.* (pp 143–174) Washington, DC: APA Books.

Pollitt, E. (1988). A critical view of three decades of research on the effects of chronic malnutrition on behavioral development. In B. Schurch & N. Scrimshaw (Eds.), *Chronic energy deficiency: Consequences and related issues.* Lausanne: IDECG.

Pollitt, E., Gorman, K., Engle, P., Martorell, R., & Rivera, J. (1993). Early supplementary feeding and cognition: Effects over two decades. *Monographs of the Society for Research in Child Development, 58* (7).

Ramey, C., & Campbell, F. (1987). The Carolina Abecedarian Project: An educational experiment concerning human malleability. In J. Gallagher & C. Ramey, (Eds.), *The malleability of children* (pp. 127–140). Baltimore: Brookes.
Richman, A., Miller, P., & LeVine, R. (1992). Culture and educational variation in maternal responsiveness. *Developmental Psychology, 28*, 614–621.
Richter, L. (1989). Household density, family size and the growth and development of Black children. *South African Journal of Psychology, 19*, 191–198.
Rogoff, B. (1990). *Apprenticeship in thinking: Cognitive development in social context.* New York: Oxford University Press.
Rogoff, B., Mistry, J., Goncu, A., Mosier, C. (1991). Cultural variation in the role relations of toddlers and their families. In M. Bornstein (Ed.), *Cultural approaches to parenting.* (pp 173–184). Hillsdale NJ: Erlbaum.
Rovee-Collier, C. (1984). The ontogeny of learning and memory in human infancy. In R. Kail & N. Spear (Eds.), *Comparative perspectives on the development of memory* (pp. 103–134). Hillsdale: Erlbaum.
Rutter, M. (1983). Statistical and personal interactions. In D. Magnusson & V. Allen (Eds.), *Human development: An interactional perspective.* (pp 295–320) New York: Academic Press.
Saco-Pollitt, C., & Pollitt, E. (1994). *Biophysical environmental factors.* Manuscript submitted for publication.
Scarr, S. (1985). Constructing psychology. *American Psychologist, 40*, 499–512.
Scarr, S. (1992). Developmental theories of the 1990s. *Child Development, 63*, 1–19.
Sigman, M., Neumann, C., Jansen, A., & Bwibo, N. (1989). Cognitive ability of Kenyan children in relation to nutrition, family characteristics and education. *Child Development, 60*, 1463–1474.
Simeon, D., & Grantham-McGregor, S. (1990). Nutritional deficiencies and children's behavior and mental development. *Nutrition Research Review, 3*, 1–24.
Steinberg, L., Dornbusch, S., & Brown, B. (1992). Ethnic differences in adolescent achievement. *American Psychologist, 47*, 723–729.
Steinberg, L., Mount, T., Lanborn, S., & Dornbusch, S. (1991). Authoritative parenting and adolescent adjustment across varied ecological niches. *Journal of Research on Adolescence, 1*, 19–36.
Stevenson, H. (1982). Influence of schooling on cognitive development. In D. Wagner & H. Stevenson (Eds.), *Cultural perspectives in child development* (pp. 208–224). San Francisco: Freemont.
Stevenson, H., & Lee, S. (1990). Contexts of achievement. *Monographs of the Society for Research in Child Development, 55*, Serial #221.
Super, C., & Harkness, S. (1986). The developmental niche. A conceptualization at the interface of child and culture. *International Journal of Behavioral Development, 9*, 545–569.
Super, C., Clement, J., Vuori, L., Christiansen, N., Mora, J., & Herrera, M. (1981). Infant and caretaker behavior as mediators of nutritional and social intervention in the barrios of Boguta. In T. Field, A. Sostek, P. Vietze, & P. Leiderman (Eds.), *Culture and early interaction* (pp. 171–188). Hillsdale NJ: Erlbaum.
Wachs, T. D. (1989). The nature of the physical microenvironment: An expanded classification system. *Merrill Palmer Quarterly, 35*, 399–402.

Wachs, T. D. (1992). *The nature of nurture.* Newbury Park CA: Sage.

Wachs, T. D. (1993a). Nature of relations between the physical and social microenvironment of the two year child. *Early Development and Parenting, 2,* 81–88.

Wachs, T. D. (1993b). The nature–nurture gap: What we have is a failure to collaborate. In R. Plomin & G. McLearn (Eds.), *Nature, nurture, and psychology.* (pp. 375–394) Washington DC: American Psychological Association.

Wachs, T. D., Bishry, Z., Sobhy, A., McCabe, G., Sheehan, F., & Galal, O. (1993). Relation of rearing environment to adaptive behavior of Egyptian toddlers. *Child Development, 64,* 586–604.

Wachs, T. D., & Camli, O. (1991). Do ecological or individual characteristics mediate the influence of the physical environment upon mother infant transactions? *Journal of Environmental Psychology, 11,* 249–264.

Wachs, T. D., & Gandour, M. (1983). Temperament, environment, and six month cognitive–intellectual development. *International Journal of Behavioral Development, 6,* 135–152.

Wachs, T. D., & Gruen, G. (1982). *Early experience and human development.* New York: John Wiley and Sons.

Wachs, T. D., Moussa, W., Bishry, Z., Yunis, S., Sobhy, A., McCabe, G., Jerome, N., Galal, O., Harrison, G., & Kirksey, A. (1993). Relations between nutrition and cognitive performance in Egyptian toddlers. *Intelligence, 17,* 151–172.

Wachs, T. D., & Plomin, R. (Eds). (1991). *Conceptualization and measurement of organism–environment interaction.* Washington, DC: American Psychological Association.

Whiting, J. (1981). Environmental constraints on infant care practices. In R. Monroe, R. Monroe, & B. Whiting (Eds.), *Handbook of cross cultural human development* (pp. 155–179). New York: Garland.

Whiting, G., & Edwards, C. (1988). *Children of different worlds.* Cambridge: Harvard University Press.

Willerman, L. (1979). The effects of families on intellectual development. *American Psychologist, 34,* 923–929.

Wohlwill, J., & Heft, H. (1987). The physical environment and the development of the child. In I. Altman & J. Stokols (Eds.), *Handbook of environmental psychology* (pp. 281–328). New York: John Wiley and Sons.

Chapter 7

How to Get to Carnegie Hall: Implications of Exceptional Performance for Understanding Environmental Influences on Intelligence

Richard K. Wagner
William L. Oliver

Florida State University

Tourist to New York City cab driver: "How do I get to Carnegie Hall?"
Cab driver to tourist: "Practice, man, practice!"

Most everyone has enough of an appreciation of the range of ordinary levels of performance to recognize and admire extraordinary performance when they see it. Thus, weekend golfers are drawn to the television when a PGA tournament is broadcast, just as mediocre tennis players (like the present authors) are drawn to Wimbledon Championship broadcasts.

This kind of recognition and admiration extends to intellectual tasks as well. Consider the common digit span task, found on most individually administered IQ tests. The typical span for digits recalled forward is under 10, yet exceptionally skilled individuals have recorded digit spans in excess of 100 digits (Chase & Ericsson, 1981; Ericsson, 1988; Staszewski, 1987). Assuming 7 plus or minus 2 as ballpark estimates of mean and standard deviation for digit span, a span of 100 digits corresponds to an IQ of approximately 800, or an SAT score of approximately 5300 (possibly higher, if we figure in completion of a Kaplan's "Preparation for the SAT" course).

Of course, the digit span task involves recall of random sequences of digits presented orally. When the task is to memorize a single string of digits that may be studied repeatedly, performance improves noticeably. Consider the case of *pi*. For most of us, our knowledge of the value of *pi* is limited to 3.1416. Yet a graduate student in our cognitive psychology program has memorized *pi* to 30,000 places, and the world record approaches 100,000 places.

What accounts for this level of extraordinary memory performance, the skill exhibited by the chess master who defeats an entire room full of skilled players when playing them simultaneously, or the mental calculation performance of an individual who can solve complex arithmetic calculations faster than engineers in the audience using calculators? More importantly, what are the implications, if any, of the development of extraordinary performance for our understanding of environmental influences on intelligence? The vast majority of intervention programs designed to increase levels of cognitive ability (e.g., Head Start) or academic achievement (e.g., Chapter I reading or math) have been modest in intensity, and length of participation has exceeded several years only rarely. In contrast, the programs that we focus on here involve training regimes followed by world-class performers that are remarkably high in intensity (typically a minimum of several hours of intense, exhausting effort per day) and in duration (typically a decade or more).

Our major goal in this chapter is to describe the development of extraordinary proficiency under admittedly extreme conditions, and then to consider evidence that one or more of the key aspects of the development of extraordinary proficiency generalizes to more ordinary environmental influences on the development of intelligence.

THE DEVELOPMENT OF EXTRAORDINARY PROFICIENCY

The common assumption, particularly among individuals other than developmental psychologists and behavior geneticists, is that the development of extraordinary levels of proficiency depends largely on innate, genetically transmitted characteristics. Thus, in the latter half of the nineteenth century, Galton (1869/1979) argued that eminence was largely due to inherited natural ability, the development of which was analogous to the development of physical skill:

> So long as he is a novice, he perhaps flatters himself there is hardly an assignable limit to the education of his muscles; but the daily gain is soon discovered to diminish, and at last it vanishes altogether. His maximum performance becomes a rigidly determinate quantity. (cited in Ericsson, Krampe, &Tesch-Römer, 1993)

Yet, predating Galton and even perhaps the joke about how to get to Carnegie Hall is an alternative account by Plato:

> What I assert is that every man who is going to be good at any pursuit must practice that special pursuit from infancy, by using all the implements of his pursuit both in his play and in his work. For example, the man who is to make a good builder must play at building houses, and to make a good farmer he must play at tilling land; and those who are rearing them must provide each child with toy tools modeled on real ones. Besides this, they ought to have elementary instruction in all the necessary subjects—the carpenter, for instance, being taught in play the use of rule and measure, the soldier being taught riding or some similar accomplishment. So, by means of their games, we should endeavor to turn the tastes and desires of the children in the direction of that object which forms their ultimate goal. (*The Laws*, Book I, cited in Wachs, 1992)

Modern behavioral genetics suggests that neither of these views is likely to be completely correct. Intellectual proficiency and other behavioral characteristics are the culmination of long-term, ongoing joint actions and probable complex interactions of genetic and environmental influences (Plomin & Bergman, 1991; Wachs, 1992).

Although areas of extraordinary proficiency appear to be domain-specific, a comparison of the development of extraordinary proficiency in a variety of domains has served to identify general principles that apply to the development of extraordinary proficiency regardless of domain (Ericsson et al., 1993; Ericsson & Smith, 1991). Our concern here, however, is whether one or more of these general principles can be applied to the case of more ordinary environmental influences on the broader array of cognitive abilities that make up intelligence.

Principles Underlying the Development of Extraordinary Levels of Proficiency

The study of the development of extraordinary performance is a fascinating field that we cannot do justice to in the space available. We will thus limit our remarks to providing a brief description of four key principles that arise from studies of the development of extraordinary proficiency. We have drawn heavily from a recent theoretical account provided by Ericsson et al. (1993), and strongly encourage the interested reader to consult this fascinating source directly.

Experts are made, not born. Despite the common belief that many experts were child prodigies who acquired their amazing skills with little effort, the evidence suggests that experts are made, not born. For example, in their classic study of chess, Simon and Chase (1973) noted that no one had attained the level of international grand master with less than about a decade of intensive study. Bobby Fischer, perhaps regarded as the best example of a modern chess prodigy, required at least nine years of preparation (Krogius, 1976). Hayes (1981) reported similar findings from the field of music. Eminent composers who produced distinguished work as young as their early twenties had begun their music training roughly 15 years previously. The finding that top levels of performance require at least a decade of training has now been confirmed for a wide variety of domains, such as music (Sosniak, 1985), tennis (Monsaas, 1985), science and business (Simonton, 1988), and running (Wallingford, 1975).

The fact that a decade or more of training is required for the attainment of top levels of performance does not mean that everyone is equally capable of such attainment, or even that individuals profit equally from training. And it is possible that individual differences in skill levels exhibited by young novices play a role in selection for training to become an expert. The magnitude of these individual differences in skill levels of novice performers, however, pale in comparison to the differences in the ultimate levels of performance attained as a function of whether sustained training is undergone. Training plays a major role in the accomplishments of even the most gifted individuals. Experts are not born demonstrating world-class levels of performance right out of the blocks; rather, they are made through a decade or more of training.

All practice is not equally profitable. If all that were required to attain world-class levels of performance was a decade of experience, the authors of this chapter would be having strawberries and cream after completing our matches at Wimbleton. The sad fact that we are not doing so suggests that there is more to the development of extraordinary proficiency than mere long-term experience.

Studies of the development of expertise suggest a key role for deliberate practice, which typically consists of a coached regime of intense and effortful practice activities designed for the express purpose of improving the current level of performance (Ericsson & Krampe, in press). In fact, differences in the estimated cumulative amount of deliberate practice distinguish top-level from second-tier performers.

Expertise isn't cheap. Because extraordinary levels of proficiency require a decade or more of training, and that such training is specialized, typically coached, and intensive, the development of expertise is costly for both society and for individuals who seek to become experts. Three kinds of constraints limit the development of expertise to selected members of society (Ericsson et al., 1993). The first is a constraint on resources. World-class performers typically begin their training regime in early childhood, initially supported and encouraged by parents, and later by coaches, as well. The time and money required to support a child on this kind of fast track are considerable. Parents must be willing to transport their children to regional and even national competitions. Depending on the domain, it even may be necessary to move to be in close proximity to the best coaching and highest level of competition. The constraint on resources is especially apparent in academic disciplines; many students who apply to graduate programs are rejected. The second kind of constraint is one on effort. We may daydream about playing professional sports, becoming a grand master in chess, or becoming the most prolific and widely cited researcher in one's field, but most of us will never live out our daydreams because the effort required involves adding several hours of intense, daily training to our already crowded days. The third kind of constraint is one on motivation. Most of us lack the motivation to seek and endure a decade or more of intense activity that is only occasionally rewarding. For every individual who makes it, countless others will burn out and cease training.

Knowledge is power. Experts acquire huge amounts of domain-specific knowledge, including knowledge about effective strategies, and it is this knowledge rather than unusual ability that accounts for their extraordinary proficiency (Charness, 1991; Patel & Groen, 1986; Sloboda, 1985). Evidence for this assertion is provided by the superior memory and pattern recognition abilities of experts in their given domains, and by the domain-specific nature of extraordinary proficiency itself.

With the development of knowledge about effective strategies, ways in which tasks are accomplished become transformed over time in ways that overcome processing limitations of novices. For example, "hunt-and-peck" sequential typing gives way to a more simultaneous process of touch typing (Salthouse, 1986), and extraordinarily long digit spans are made possible for an individual who is a runner by remembering sequences of digits as running times for races of various durations (Chase & Ericsson, 1981).

GENERALIZATION TO THE CASE OF ENVIRONMENTAL INFLUENCES ON INTELLIGENCE

Obvious differences exist between the development of extraordinary levels of accomplishment through extreme interventions in the form of training regimes undergone by top performers and the more general case of environmental influences on intelligence. Perhaps the most salient is that extraordinary proficiency appears to be constrained tightly to a given domain, whereas we think of intelligence as representing a much broader kind of intellectual proficiency. We don't, for example, recruit top chess masters to be generals in the armed services, because we do not believe that their keen ability to foresee and counter moves by an opponent on a chess board represents a general ability to outguess opponents that will be demonstrated equally well on the battlefield. A second difference is that given the resources required, society cannot possibly afford to develop expertise as universally as it attempts to develop broader intellectual functioning through schooling.

Nevertheless, a fundamental issue that has motivated study of the development of extraordinary proficiency is the same one that has been at the heart of the study of environmental influences on intelligence; namely, does proficiency primarily reflect innate, genetically-transmitted ability or experience? Might key principles derived from the study of the development of extraordinary proficiency have their counterparts in the case of environmental influences on intelligence?

Because one's conclusions about environmental influences on intelligence depend monumentally on what one means by intelligence and by environmental influences, we begin by sketching the conceptualizations of these two constructs that we have adopted for the present purpose. Then, we turn to our main topic of how lessons learned from the study of the development of extraordinary performance generalize to the case of environmental influences on intelligence, considering in turn (a) changing relations between ability and performance with practice; (b) the effects of short-term practice on simple cognitive tasks, using the example of simple reaction time tasks that purport to be measures of basic cognitive processes; and (c) the effects of long-term practice on more complex tasks, using the example of the effects of schooling on intellectual functioning.

Conceptualizing Intelligence and Environments

The task of considering environmental influences on intelligence would be hard enough were there general agreement about what is meant by intelligence and by environments. Answers to the questions of how, and how much is intelligence influenced by the environment, are likely to be different for different conceptualizations of intelligence and environments (Messick & Sigel, 1982; Wachs, 1992). Unfortunately, remarkably little consensus exists around either construct. Although undesirable, this state of affairs need not be paralyzing, provided we put our cards on the table by being specific about our conceptualizations of intelligence and environments, thereby providing the reader with the context necessary for evaluating our subsequent remarks.

Conceptualizing intelligence. When 17 leading researchers were asked to define intelligence in a symposium sponsored by the *Journal of Educational Psychology* in 1921, the response was 14 different definitions and 3 nonreplies.

A second attempt 65 years later showed little, if any more, consensus among contemporary replies (Sternberg & Detterman, 1986). Of the many issues surrounding the conceptualization of intelligence, three are particularly relevant in the present context.

Complexities in distilling genotypic behavior from phenotypic intelligence. It goes without saying that IQ, *g* (the general factor that represents common variance among cognitive tasks), simple or choice reaction time, or any other behaviorally-based measure that we might propose as an index of intelligence represent the culmination of long-term, ongoing joint actions and probable complex interactions of genetic and environmental influences. At the present time, no behavioral measures provide direct indices of what we might call "biological" intelligence, and it is unlikely that this situation will change in the foreseeable future. For example, the apparent simplicity of some reaction-time tasks has been used to argue that such tasks reflect basic elementary processes, or even differences in the speed and accuracy of neural transmission (Eysenck, 1982; Vernon, 1987). Yet large differences in performance on such tasks occur with relatively minor changes in task instructions and apparatus (Detterman, 1987; Detterman & Andrist, 1990). Similarly, the substantial heritabilities reported for IQ (perhaps on the order of .50) should not mistakenly be the basis for concluding that IQ, or *g*, provides a rough index of variation in intelligence that is of genetic origin.

The complexities involved in studying relations between genetic and environmental causes are actually enormous. These complexities are characteristic of the data, as well as their interpretation and analysis. For an example of complexities of data, heritability of IQ appears to vary as a function of level. Groups of low-IQ individuals show higher heritability and less shared environmental influences than do groups of high-IQ individuals (Detterman, Thompson, & Plomin, 1989). In fact, similar differences even exist for correlations among cognitive ability tests, and consequently, the proportion of variance accounted for by a first factor (*g*). These correlations are quite a bit larger for individuals who score low on IQ tests than for individuals who score high (Detterman & Daniel, 1989).

As an example of complexities of data interpretation and analysis, structural equation based multivariate models provide simultaneous, maximum-likelihood parameter estimates for genetic and environmental causes, a marked advance over previous methods for analyzing twin and other kinds of studies (Plomin, DeFries, & McClearn, 1990)—yet worrisome issues remain. Models that seek to partition variance into genetic and environmental sources typically assume additive effects of genetic and environmental influences, which (a) ignores the probable interaction or covariance between genetic and environmental influences, (b) can yield "impossible" parameter estimates such as negative variances, and (c) does not inform the underlying processes by which genotypes become phenotypes (Bronfenbrenner, 1991; Schonemann & Schonemann, 1991; Wachs, 1991; Wahlsten, 1990). Results from twins and adoptive families are assumed to generalize to the majority of families that have neither, despite the possibility that parents may treat twins and adoptees differently than non-twin biological offspring (Baumrind, 1991). Finally, behavioral genetic studies of human behavior are correlational, but experimental studies are the norm for genetic studies of other species. The experimental manipulation is assigned mating. Heritabilities derived from experimental studies with animals, even for physical traits, typically are considerably less than heritability estimates derived from human correlational studies (Scott, 1987). For example, heritabilities for milk production in cattle are about 30%, and about 25% for staple length of wool in sheep (King, 1981, cited in Schonemann & Schonemann, 1991).

We have no doubt that there are genetic influences on common observable indices of intelligence, but there are environmental influences and probable genetic–environmental interactive influences, as well. The bottom line is that the best we can do is sample phenotypic behavior, and recognize that attempting to distill genotypic behavior from phenotypic behavior requires due consideration to attendant assumptions and their verification in any given situation.

Aptitude versus achievement. If we acknowledge the complexities of trying to distill genotypes from phenotypes, can we at least assume that common aptitude tests indeed measure aptitude for learning, as opposed to merely sampling what has been learned, or achievement? Our view is that IQ and aptitude tests do not measure aptitude for learning, although it is possible to make *inferences* about aptitude from performance on such tasks if one is willing and able to support several key assumptions; namely, equal (a) opportunity to learn the information and to acquire the task strategies that are tested, (b) motivation to perform, and (c) familiarity with the testing materials and context. Such assumptions may be tenable for homogeneous samples of young children, but becomes increasingly untenable for adults as they sort themselves into more varied environments upon completion of their formal schooling.

Rank-ordering versus levels of performance. The vast majority of research on intelligence has concerned individual differences, or the rank-orders of individuals. Much less research has focused on changes in overall levels of performance, as a consequence, say, of development or changes in society over the decades. The distinction between improving average levels of performance on the one hand, and changing a rank ordering of performance on the other, has implications for attempts to make basic changes in intellectual performance. Enrichment and training programs that would be judged to be a success on the basis of gains in levels of performance are likely to be judged to be a failure if the criterion is change in rank-ordering on a cognitive measure (Brown & Campione, 1982). Changes in rank order are difficult to bring about when a program is provided to both high and low performers as opposed only to low performers, and when there are not "ceiling effects" either on what is trained or what is tested upon completion of training.

Conceptualizing environments. Conceptualizing environments used to be easy, in that environments used to be defined by SES (socioeconomic status); the numbers of things in the home such as books, siblings and parents (biological versus otherwise); the amount of time spent watching television; whether there were one or two wage earners in the home; whether children stayed at home when young or attended nursery and preschools; and so on. But these conceptualizations turn out to be gross oversimplifications.

Easily tabulated and observed characteristics of environments provide an incomplete picture of environmental influences. Individuals do not passively experience their environments; rather, they play a major role in creating their own environments, bringing to bear their intellect and prior knowledge, interests, and motivations (Scarr & McCartney, 1983). The environment of the cockpit of a Boeing 747 as experienced by a pilot who flies this type of airplane bears little resemblance to the experience of being in the same cockpit for those of us who are not pilots. And because environments are not apprehended directly, but rather perceived and filtered through our own eyes, ears, and prior experience, seemingly environmental variables such as SES can represent genetic as well as environmental influences (Plomin & Bergman, 1991). For example, SES may influence intellectual development, but intellectual development, in turn, may be one of many influences that determine SES. In summary, environments are best conceptualized as complex, interrelated entities and are mediated and, indeed, constructed in part by individuals.

Changing Relations Between Ability and Performance

Countless validity coefficients have been reported between measures of cognitive ability and criterion performance. In general, the magnitude of these coefficients is higher for measures of school performance than for measures of out-of-school performance, and higher for performance on criterion performance measures that are test-like than for measures that are not.

In the case of the development of extraordinary proficiency, the magnitude of individual differences that underlie rank ordering of untrained performers is modest in comparison to the magnitude of differences that result from differential amounts of training. Consequently, initial status as indexed by ability measures provides less information about the probability of attaining extraordinary levels of proficiency than do environmental variables such as cumulated amount of deliberate practice (Ericsson et al., 1993). Recent analyses of relations between ability scores and performance upon completion of more typical levels of training suggest a similar phenomenon exists for ordinary levels of proficiency.

Hulin, Henry, and Noon (1990) carried out a meta-analysis of validity coefficients to assess the effects of the time interval between initial test and criterion performance on the level of prediction provided. The predictors ranged from IQ scores to measures of physical coordination, and the performance domains ranged from college, to graduate school, and to occupations and athletics. From initial performance to final performance, the average decrement in validity coefficient was −.6, after correction for range restriction and unreliability. This means that measures of ability were predictive of initial levels of performance, but increasingly unpredictive of performance as it develops over time. In fact, whether criterion performance itself is poor or excellent at an early stage of learning may not be very predictive of eventual levels of performance. Ackerman (1987, 1988) made a similar observation about differential relations between ability and performance as a function of level of skill acquisition. Early performance is related to general cognitive ability, whereas later performance is more related to perceptual–motor abilities.

The origin of changing relations between ability and performance with skill development is unclear. Part of the answer is probably changes in the nature of task performance with practice, as initial resource-dependent, controlled processing gives way to increasingly resource-independent, automatic processing (Ackerman, 1987; Anderson, 1982; Salthouse, 1991). According to this view, skill development results in circumventing human processing limitations evident during initial levels of performance. For example, the superior memory performance of experts appears to be the result of chunking and rapid, effective storage in long-term memory (Charness, 1976; Ericsson & Staszewski, 1989). The end result is overcoming the usual short-term memory processing limitations. An alternative view of the origin of changing relations between ability and performance with skill development is that ability is not fixed, and undergoes improvement when proficiency is acquired in complex tasks (Hulin, Henry, & Noon, 1990).

In summary, the results of research into changing relations between ability and performance with skill development parallel those observed in the study of the extraordinary skill development of world-class performers. In both cases, ultimate levels of proficiency are not well predicted by ability before training for skill acquisition has begun.

Effects of Short-Term Practice on Simple Cognitive Tasks

We noted previously the argument that a correlation between performance on simple reaction time (RT) tasks and IQ has been used to suggest that the correlation reflects basic information processing efficiency or, perhaps, even the speed and accuracy of neural transmission (Eysenck, 1982; Vernon, 1987). Because of the apparent simplicity of many reaction time tasks, it has been argued that such tasks "show no learning, do not involve problem solving, and are not dependent in any sense on previous learning. That they nevertheless correlate very highly with IQ tests must throw serious doubt on (environmental theories of intelligence)" (Eysenck, 1988, p. 92, cited in Ceci, 1990).

However, the ubiquitous finding from countless studies is that task performance on simple, as well as complex, tasks improves with practice (see, e.g.,

Longstreth, 1984; Razz, Wilerman, Igmundsen, & Hanlon, 1983). In fact, theories have been devised to account for improvement in task performance with practice (Anderson, 1982; Fitts & Posner, 1967; Schneider & Shiffren, 1977; Shiffren & Schneider; 1977). In addition, previous learning in the form of differential familiarity with the task stimuli used also affects performance on simple cognitive tasks (Tetewsky, 1992).

For simple cognitive tasks, the improvement of performance with practice probably represents increasing automaticity of components of task performance, thereby circumventing processing limitations. Salthouse (1984, 1991) studied the case of the development of skill at transcription typing. Given the fact that reaction times to successively presented stimuli are about .5 seconds, the maximum projected rate, if each letter is processed sequentially when typing, is on the order of 20 words per minute. Yet highly skilled typists perform at well over 4 times this rate. They appear to do so by processing up to 7 characters ahead of the letter being typed, and by executing motor movements simultaneously using overlapping movement. For example, Salthouse compared the speed with which two types of letter sequences—typing the same letter twice (e.g., "ff"), and typing a pair of letters that require use of both hands ("fj")—were processed by highly-skilled and less-skilled typists. The results were a disordinal interaction between skill at typing and letter sequence type. Less-skilled typists were faster at typing the same letter twice than they were letter pairs that required using both hands. Highly-skilled typists showed the opposite pattern, being faster on letter pairs that required the use of both hands. What the highly-skilled typists apparently were able to do to a greater degree than the less-skilled typists was to initiate motor movements simultaneously, by beginning the required finger movement of the second hand before completing the finger movement of the first hand.

In summary, performance on even simple cognitive tasks appears to be related to individual differences in practice, familiarity, and motivation (Ceci, 1990). As was the case for the development of extraordinary proficiency, performance on simple cognitive tasks improves with practice, and part of the improvement appears to result from the ability to restructure task requirements and strategies as skill is developed.

Effects of Long-Term Practice on Complex Cognitive Tasks: The Case of Schooling

Fred Morrison compared Canadian children whose birthdates fell just after the cut-off for school entrance with those whose birthdates fell just before the cut-off and followed two groups of children for a total of two years, who differed in average age only by a couple of months, but by an entire year of schooling. The results showed that the additional schooling resulted in improved memory performance and strategy usage.

In general, there is a good deal of evidence of relations between schooling and IQ that, when taken together, suggest that schooling influences IQ, although it is likely that IQ also affects schooling variables as well. Ceci (1990) made this case by noting that increments in IQ are associated with (a) total years of schooling (at a level of .7); (b) amount of schooling completed by same-aged, same-SES children; (c) increasing mean numbers of years of schooling completed due to intergenerational changes in schooling; and (d) Blacks moving from the South to better schools in the North. Conversely, decrements in IQ are associated with (a) summer vacation; (b) delayed commencement of schooling; and (c) early termination of schooling.

In summary, the kind of long-term practice provided by schooling appears to influence the development of subsequent intelligence. An interesting similarity, which concerns expense and resource constraints, exists between the development of extraordinary proficiency and the development of intelligence through schooling. The training regimes undergone by world-class performers typically are begun as children, requiring the support of parents and coaches. Training often consists of several hours of supervised practice daily, carried out for a decade or more. These characteristics also applied to schooling in modern societies. Generations ago, children were likely to enter the fields, and, later, the factories before completion of schooling—families could not afford to do without child labor. Now, however, an increased standard of living and changing values afford an opportunity for children to do little more than attend school for 12 or more years, akin to the conditions required to support the development of extraordinary proficiency. Teachers serve in the role of coaches, and, in basic subjects, training exercises are provided with the goal of improving current skill levels. The result is remarkable development in reading and math for most children over the course of their schooling.

In contrast to the learning of reading and mathematics skills, much of the learning in school requires students to learn facts for the purpose of passing tests. The information that is learned is not integrated into existing skills or used in subsequent learning episodes. The predominance of this type of learning in school settings, particularly at universities, may help explain puzzling findings from a number of studies showing virtually no relation between the amount of time students study and the grades they receive in their courses (e.g., Schuman, Walsh, Olson, & Etheridge, 1985). In these studies, standardized tests do the best job of predicting course grades, suggesting that they tap into the study skills needed to do well on tests of factual knowledge—such skills would include cramming (it has been estimated that more than 40 percent of students chronically cram for exams; Rothblum, Solomon, & Murakami, 1986) and shrewd guessing on multiple choice exams. This interpretation raises the concern, voiced by many and beyond the scope of this paper, that schooling fosters skills that are primarily useful for passing course exams and standardized tests.

CONCLUSIONS

Although there are obvious differences between the development of extraordinary proficiency and more typical environmental influences on intelligence, principles that characterize the development of extraordinary proficiency appear to apply to the case of environmental influences on intelligence. Measures of ability are predictive of initial performance, but are less predictive of the eventual level of performance attained compared to characteristics of the training environment, such as cumulative amounts of effective training. Practice and learning history affect performance on both simple and complex cognitive tasks.

Sustained training is required for the development of skilled performance, regardless of apparent differences in ability. This is not to say that everyone can achieve skilled performance in any domain with training, nor that individuals profit from training to the same degree. The development of skilled performance, and of the broader repertoire of cognitive skills, knowledge, and strategies that make up intelligence, represent the culmination of the actions, and interactions, of genetic and environmental influences over a prolonged developmental period. Study of development of extraordinary proficiency provides a vehicle for examining developmental processes that potentially underlie the development of intelligence.

REFERENCES

Ackerman, P. L. (1987). Individual differences in skill learning: An integration of psychometric and information processing perspectives. *Psychological Bulletin, 102*, 3–27.

Ackerman, P. L. (1988). Determinants of individual differences during skill acquisition: Cognitive abilities and information processing. *Journal of Experimental Psychology: General, 117*, 288–318.

Anderson, J. R. (1982). Acquisition of cognitive skill. *Psychological Review, 89*, 369–406.

Baumrind, D. (1991). To nurture nature. *Behavioral and Brain Sciences, 14*, 386–387.

Bronfenbrenner, U. (1991). The nature of nurture. *Behavioral and Brain Sciences, 14*, 390–391.

Brown, A. L., & Campione, J. C. (1982). Modifying intelligence or modifying cognitive skills: More than a semantic quibble? In D. K. Detterman & R. J. Sternberg (Eds.), *How and how much can intelligence be increased* (pp. 215–230). Norwood, NJ: Ablex.

Ceci, S. J. (1990). *On intelligence . . . more or less: A bio-ecological treatise on intellectual development*. Englewood Cliffs, NJ: Prentice Hall.

Charness, N. (1976). Memory for chess positions: Resistance to interference. *Journal of Experimental Psychology: Human Learning and Memory, 2*, 641–653.

Charness, N. (1991). Expertise in chess: The balance between knowledge and search. In K. A. Ericsson & J. Smith (Eds.), *Toward a general theory of expertise: Prospects and limits* (pp. 39–63). New York: Cambridge.

Chase, W. G., & Ericsson, K. A. (1981). Skilled memory. In J. R. Anderson (Ed.), *Cognitive skills and their acquisition* (pp. 141–189). Hillsdale, NJ: Erlbaum.

Detterman, D. K. (1987). What does reaction time tell us about intelligence? In P. A. Vernon (Ed.), *Speed of information-processing and intelligence* (pp. 177–199). Norwood, NJ: Ablex.

Detterman, D. K., & Andrist, C. G. (1990). Effect of instructions on elementary cognitive tasks sensitive to individual differences. *American Journal of Psychology, 103*, 367–390.

Detterman, D. K., & Daniel, M. H. (1989). Correlations of mental tests with each other and with cognitive variables are highest for low IQ groups. *Intelligence, 13*, 349–359.

Detterman, D. K., Thompson, L. A., & Plomin, R. (1989). Differences in heritability across groups differing in ability. *Behavior Genetics, 20*, 369–384.

Ericsson, K. A. (1988). Analysis of memory performance in terms of memory skill. In R. J. Sternberg (Ed.), *Advances in the psychology of human intelligence, Vol. 5* (pp. 137–179). Hillsdale, NJ: Erlbaum.

Ericsson, K. A., Krampe, R. T., & Tesch-Römer, C. (1993). The role of deliberate practice in the acquisition of expert performance. *Psychological Review, 100*, 363–406.

Ericsson, K. A., & Smith, J. (Eds.) (1991). *Toward a general theory of expertise*. New York: Cambridge University Press.

Ericsson, K. A., & Staszewski, J. (1989). Skilled memory and expertise. In D. Klahr & K. Kotovsky (Eds.), *Complex information processing* (pp. 235–267). Hillsdale, NJ: Erlbaum.

Eysenck, H. J. (Ed.) (1982). *A model for intelligence*. New York: Springer-Verlag.

Eysenck, H. J. (1988). The biological basis of intelligence. In S. H. Irvine & J. W. Berry (Eds.), *Human abilities in cultural context* (pp. 87–104). New York: Cambridge University Press.

Fitts, P., & Posner, M. I. (1967). *Human performance*. Belmont, CA: Brooks/Cole.

Galton, F., Sir (1869/1979). *Hereditary genius: An inquiry into its laws and consequences*. London: Julian Friedman Publishers.

Hayes, J. R. (1981). *The complete problem solver*. Philadelphia, PA: Franklin Institute Press.

Hulin, C. L., Henry, R. A., & Noon, S. L. (1990). Adding a dimension: Time as a factor in the generalizability of predictive relationships. *Psychological Bulletin, 107*, 328–340.

King, J. C. (1981). *The biology of race*. University of California Press.

Krogius, N. (1976). *Psychology in chess*. New York: RHM Press.

Longstreth, L. E. (1984). Jensen's reaction-time investigations of intelligence: A critique. *Intelligence, 8*, 139–160.

Messick, S., & Sigel, I. (1982). Conceptual and methodological issues in facilitating growth and intelligence. In D. K. Detterman & R. J. Sternberg (Eds.), *How and how much can intelligence be increased* (pp. 187–195). Norwood, NJ: Ablex.

Monsaas, J. A. (1985). Learning to be a world-class tennis player. In B. S. Bloom (Ed.), *Developing talent in young people* (pp. 211–269). New York: Ballantine Books.

Patel, V. L., & Groen, G. J. (1986). Knowledge based solution strategies in medical reasoning. *Cognitive Science, 10*, 91–116.

Plomin, R., & Bergman, C. S. (1991). *Behavioral and Brain Sciences, 14*, 373–427.

Plomin, R., DeFries, J. C., & McClearn, G. E. (1990). *Behavioral genetics: A primer* (2nd. ed.). New York: W. H. Freeman.

Razz, N., Wilerman, L., Igmundsen, P., & Hanlon, M. (1983). Aptitude-related differences in auditory recognition masking. *Intelligence, 7*, 71–90.

Rothblum, E. D., Solomon, L. J., & Murakami, J. (1986). Affective, cognitive, and behavioral differences between high and low procrastinators. *Journal of Counseling Psychology, 33*, 387–394.

Salthouse, T. A. (1984). Effects of age and skill in typing. *Journal of Experimental Psychology: General, 117*, 115–129.

Salthouse, T. (1986). Perceptual, cognitive, and motoric aspects of transcription typing. *Psychological Bulletin, 99*, 303–319.

Salthouse, T. A. (1991). Expertise as the circumvention of human processing limitations. In K. A. Ericsson & J. Smith (Eds.), *Toward a general theory of expertise* (pp. 286–300). New York: Cambridge University Press.

Scarr, S., & McCartney, K. (1983). How people make their own environments: A theory of genotype to environmental effects. *Child Development, 54*, 424–435.

Schneider, W., & Shiffrin, R. M. (1977). Controlled and automatic information processing: I. Detection, search, and attention. *Psychological Review, 84*, 1–66.

Schonemann, P. H., & Schonemann, R. D. (1991). We wondered where the errors went. *Behavioral and Brain Sciences, 14*, 404–406.

Schuman, H., Walsh, E., Olson, C., & Etheridge, B. (1985). Effort and reward: The assumption that college grades are affected by quantity of study. *Social Forces, 63*(4), 945–966.

Scott, J. P. (1987). Why does human twin research not produce results consistent with those from nonhuman species? *Behavioral and Brain Sciences, 10*, 39.

Shiffrin, R. M., & Schneider, W. (1977). Controlled and automatic information processing: II. Perceptual learning, automatic attending, and a general theory. *Psychological Review, 84*, 127–190.

Simon, H. A., & Chase, W. G. (1973). Skill in chess. *American Scientist, 61*, 394–403.

Simonton, D. K. (1988). Age and outstanding achievement: What do we know after a century of research? *Psychological Bulletin, 104*, 251–267.

Sloboda, J. A. (1985). *The musical mind: The cognitive psychology of music*. London: Oxford University Press.

Sosniak, L. A. (1985). Learning to be a concert pianist. In B. S. Bloom (Ed.), *Developing talent in young people* (pp. 19–67). New York: Ballantine Books.

Staszewski, J. J. (1987). *The psychological reality of retrieval structures: An investigation of expert knowledge*. Unpublished doctoral dissertation, Cornell University.

Sternberg, R. J., & Detterman, D. K. (1986). *What is intelligence?* Norwood, NJ: Ablex.

Tetewsky, S. J. (1992). Familiarity effects in visual comparison tasks and their implications for understanding human intelligence. *Journal of Experimental Psychology: Human Learning and Memory, 18*, 577–594.

Vernon, P. A. (Ed.) (1987). *Speed of information-processing and intelligence*. Norwood, NJ: Ablex.

Wachs, T. D. (1991). Nature and nurture: A shaky alliance. *Behavioral and Brain Sciences, 14*, 411–412.

Wachs, T. D. (1992). *The nature of nurture*. Newbury Park, CA: Sage Publications.

Wahlsten, D. (1990). Insensitivity of the analysis of variance to heredity–environment interaction. *Behavioral and Brain Sciences, 13*, 109–161.

Wallingford, R. (1975). Long distance running. In A. W. Tayler & F. Landry (Eds.), *The scientific aspects of sports training* (pp. 118–130). Springfield, IL: Charles C. Thomas.

Part 2

Commentaries on Chapters 1–7

Chapter 8

Knowledge Structures: Successive Glimpses of an Elusive Theory of Adult Intelligence

Phillip L. Ackerman

University of Minnesota

Briefly, the view expressed in this chapter is this: Adult intelligence is not simply an extension of child and adolescent intelligence. The presence of adult intellect is marked not only by what an individual can do (process), but also fundamentally by what the adult knows (knowledge structure). This chapter presents the basis for this view, in describing both historical and current views on adult intellect. In addition, an outline for future research is briefly described that focuses on assessment of intellect-as-knowledge.

BACKGROUND

Many philosophers of science often dispense with the notion of Kuhnian-type paradigms as descriptions of how science proceeds (e.g., Lakatos, 1970). Various scientists, however, have found the concept a compelling one for particular domains, even though observers disagree as to whether psychology is pre-paradigmatic or paradigmatic. Nonetheless, nearly one hundred years after the development of the Binet–Simon scales, it seems clear that intelligence theory and research is well-described as paradigm-bound. General intelligence (whether as IQ or *g*) serves as either the main focus of attention (e.g., Humphreys, 1979; Wechsler, 1944), or at the very least, the touchstone from which various departures are made (e.g., Sternberg, 1977; Thurstone, 1938). Some changes to the general construct of intelligence represent normal science, while others have the feel of Popperian bold conjectures (e.g., information processing as intelligence; simultaneous–successive processing as intelligence, etc), with the attendant ruthless refutations by the community.

However, one of the most dilatory debates about intelligence (as measured by IQ tests, information processing tests, line judgement, or whatever) during this century is the often-argued issue about intelligence being a reflection of "capacity" or "capabilities." As nearly all psychologists know by now (or *should* know), the standard IQ test measures acquired knowledge, skills, and abilities—*not* capacity (see discussions by Anastasi, 1958; Humphreys, 1979). This does not detract from the fact that practical predictions about individual differences, knowledge, and skill acquisition can be made from IQ measures (e.g., in school situations). The focus on capacity, however, is much more pernicious when dealing with the adult population. An assessment of the intelligence of an adult, who may have had most of his/her learning experiences in the past, must practically focus on what the adult currently knows and currently can do, especially when the criterion to-be-predicted is primarily a resultant of transfer-of-training rather than new learning (e.g., see Ferguson, 1954, 1956).

In recent years, though, some aspects of cognitive psychology have lifted the veil on a revolutionary approach to adult intelligence, which is far removed from the Binet–Simon scale. This approach focuses less on intelligence-as-process and more fundamentally on intelligence-as-knowledge. Some aspects of this orientation are present in this volume, most notably in the chapter by Wagner and Oliver, and also in Keating. Keating, while focusing mostly on child and adolescent intelligence, mentions the potential of treating "expertise" as a component of intellect. Wagner and Oliver, on the other hand, focus specifically on the development of exceptional performance, mostly in specific domains of expertise. The latter discussion draws on a now-extensive array of cognitive literature concerning expert knowledge in domains such as chess, music, athletics, and so on. These treatments, though, only appear to glimpse the tip of the intellectual iceberg. One can trace the modern history of this way of thinking by Binet and Simon (1905/1908), Henmon (1921), Vernon (1950), Cattell (1971/1987), and Glaser (1991), as discussed in the following section.

BINET AND SIMON (1905/1908)

By their very design, modern omnibus intelligence tests mostly assess process, as much devoid of knowledge as possible (except for what can be considered core cultural concepts, and, in some cases, information and vocabulary knowledge). This was an excellent strategy as *part* of a multi-method assessment of school-age children, as described by Binet and Simon (1905/1908):

1. *The medical method*, which aims to appreciate the anatomical, physiological, and pathological signs of inferior intelligence.
2. *The pedagogical method*, which aims to judge of the intelligence according to the sum of acquired knowledge.
3. *The psychological method*, which makes direct observations and measurements of the degree of intelligence.

Further, Binet and Simon said that:

> It is understood that we here separate natural intelligence from instruction. It is the intelligence alone that we seek to measure, by disregarding in so far as possible, the degree of instruction which the subject possesses. He should, indeed, be considered by the examiner as a complete ignoramus knowing neither how to read nor write. The necessity forces us to forego a great many exercises having a verbal, literary or scholastic character. These belong to a pedagogical examination. (trans. in Jenkins & Paterson, 1961, p 93)

The primary rationale for excluding acquired knowledge in assessment of intelligence in children, provided by Binet and Simon, was an attempt to derive an estimate of intelligence that was removed as far as possible from the advantages or disadvantages of socioeconomic status and differential exposure to school or other sources of instruction. In retrospect, for their purposes, this seems an appropriate method, and is an obviously successful one. However, as a means of assessing adult intelligence, the "pedagogical method" must become part of a full assessment of intellect.

HENMON (1921)

A popular reference point for thinking about the construct of intelligence was the compendium presented in the 1921 *Journal of Educational Psychology*. A definition proposed by Henmon seems to capture the current perspective:

> Intelligence in the ordinary acceptance of the term, has been defined by Lester F. Ward to be 'intellect coupled with the product of its operation,' or in other words,

> intelligence is intellect *plus* knowledge. . . . Intelligence, then, involves two factors—the capacity for knowledge and knowledge possessed. (Henmon, 1921, p. 195).

VERNON (1950)

As an intelligence theorist, Vernon comes closest to developing a joint representation of intelligence-as-process and intelligence-as-knowledge. He presented a schematic representation of "educational abilities" in his classic 1950

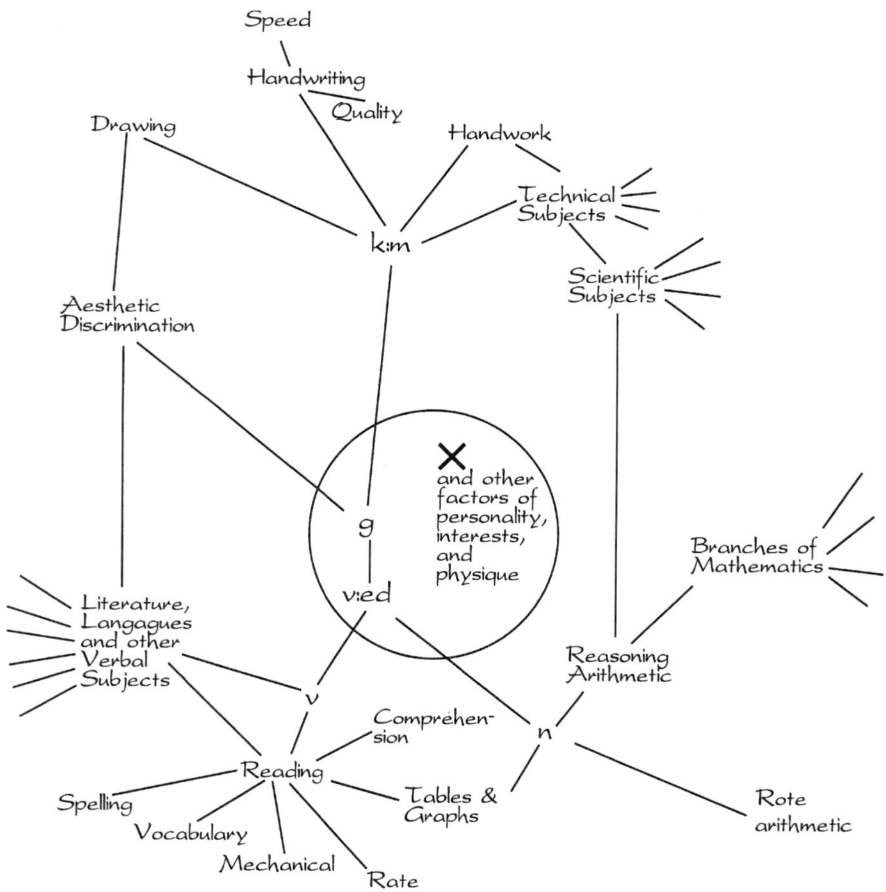

Figure 8.1. Diagram of the Structure of Educational Abilities. g = general ability; v:ed = Verbal: Educational; K:M = practical:mechanical; v = verbal; X = industriousness (including personality and interests).

Note. From *The Structure of Human Abilities* (p. 47) by P. E. Vernon, 1950, New York: John Wiley and Sons. Copyright 1950 by John Wiley and Sons. Adapted by permission.

work, shown in Figure 8.1. It should be clear from this depiction that his intent was to represent the panoply of knowledge sources found in adults (and only of limited utility for children and adolescents, who can be expected to have quite restricted knowledge in many of the domains).

CATTELL (1971/1987)

As one of the most prolific psychologists of our time, Cattell presents the reader with myriad theoretical statements about the nature of intellect, personality, and interests. Numerous intelligence theorists and empirical researchers have focused narrowly on Cattell's explication of fluid and crystallized intelligences. The Gf/Gc theory, as explicated in various articles and chapters by Cattell and Horn (e.g., Cattell, 1963; Horn, 1965), comes closest to separating intelligence-as-process (Gf) from intelligence-as-knowledge (Gc), though it must be admitted that Gc also contains several aspects of intelligence-as-process. However, less well known is the full specification of the developmental trajectory of crystallized intelligence as described by Cattell, within his "investment theory." He stated:

> However, these composite origins must be kept in mind when we begin to ask what happens to crystallized general intelligence, and the traditional intelligence tests that measure it, *after* school. The crystallized intelligence then goes awry both conceptually and in regard to the practical predictions to be made from traditional intelligence tests. In the twenty years following school, the judgmental skills that one should properly be measuring as the expression of learning by fluid ability must become different from different people. If these are sufficiently varied and lack any common core, the very concept of general intelligence begins to disappear.
>
> [The psychologist's] alternatives are then: (a) to sample behavior still more widely than in the traditional test, using a formula expressing the role of fluid intelligence in learning in each of many different fields (an approach which, in practice, might amount to producing as many different tests as there are occupations, etc.); (b) to change completely to fluid intelligence measures . . . or (c) to continue to measure by the "school version" of crystallized ability essentially learning on what the individual's intelligence was at the time of leaving school. (pp. 143–144).

A quick perusal of the aging literature suggests that psychologists have almost exclusively selected options (b) and (c), to the detriment of discovering the knowledge structures of adult intellect.

GLASER (1991)

In contrast to the earlier theorists, Glaser and his colleagues have indeed focused heavily on knowledge structures. Using artificial intelligence techniques of expert–novice examinations and large-scale learning paradigms,

Glaser has attempted to describe the knowledge contents of expertise in a few different domains (e.g., physics, economics). The work is painstaking and thorough, but is unfortunately limited to the degree that it can be applied more generally to adult intelligence in the population at large. Nonetheless, Glaser (1991) sees the future in a knowledge structure approach, as follows:

> Thus, as well as working on general theories of intelligence, we should concentrate on examining the details of knowledge use in all forms of intelligent performance. When we know what people know, we can offer more realistic theories about determinants of performance because people think largely by using quite specific knowledge. (p. 53)

A PROPOSAL

Recently (Ackerman, in press), I proposed that in the face of historical theory and empirical evidence, we can no longer expect to adequately treat adult intelligence as a simple extension of child and adolescent IQ. Theorists like Henmon, Vernon, Cattell, and Glaser have pointed the way towards a new conceptualization of adult intelligence-as-knowledge, to supplement the tradition of intelligence-as-process. For assessment of adult intelligence to proceed, we need to develop the following: (a) A taxonomy of "intellectual" knowledge; (b) test batteries that allow for assessment of a wide variety of knowledge domains; and (c) a new methodology that will translate the expected sparse matrices of breadth and depth of individuals' knowledge into real-world predictions (e.g., post-graduate school success, job performance, etc.). With development of these three components, we can hope to invigorate the study of adult intelligence, as well as provide a basis for the kind of integration among constructs of intellect-as-process, personality, and motivation that clearly interact to produce individual differences in the development and expression of intelligence. Such an integration provides a basis for evaluating all-encompassing theories like that of Cattell (1971/1987), and for development of competing models of adult development and aging effects that have important real-world implications.

REFERENCES

Ackerman, P. L. (in press). Intelligence as process and knowledge: An integration for adult development and application. In W. A. Rogers, A. D. Fisk, & N. Walker (Eds.), *Aging and skilled performance: Advances in theory and applications.* Hillsdale, NJ: Erlbaum.

Anastasi, A. (1958). Heredity, environment, and the question "How?". *Psychological Review, 65*, 197–208.

Binet, A., & Simon, T. (1905/1908). The development of intelligence in children. New methods for the diagnosis of the intellectual level of subnormals. The development of intelligence in the child. Excerpts from *L'Année Psychologique, 11*, 163–191, 191–244, and *14*, 1–90. Translated and reprinted in J. J. Jenkins & D. G. Paterson (Eds)., *Studies of individual differences: The search for intelligence*. New York: Appleton-Century-Crofts.

Cattell, R. B. (1963). Theory of fluid and crystallized intelligence: A critical experiment. *Journal of Educational Psychology, 54*, 1–22.

Cattell, R. B. (1971/1987). *Abilities: Their structure, growth and action*. [Revised and reprinted as *Intelligence: Its structure, growth, and action*]. Amsterdam: North-Holland.

Ferguson, G. A. (1954). On learning and human ability. *Canadian Journal of Psychology, 8*, 95–112.

Ferguson, G. A. (1956). On transfer and the abilities of man. *Canadian Journal of Psychology, 10*, 121–131.

Glaser, R. (1991). Intelligence as an expression of acquired knowledge. In H. A. H. Rowe (Ed.), *Intelligence: Reconceptualization and measurement* (pp. 47–56). Hillsdale, NJ: Erlbaum.

Henmon, V. A. C. (1921). Intelligence and its measurement: A symposium. *Journal of Educational Psychology, 12*, 195–198.

Horn, J. L. (1965). *Fluid and crystallized intelligence: A factor analytic study of the structure among primary mental abilities*. Unpublished doctoral dissertation, University of Illinois, Urbana.

Humphreys, L. G. (1979). The construct of general intelligence. *Intelligence, 3*, 105–120.

Keating, D. P. (in press). Habits of mind: Developmental diversity in competence and coping. In D. K. Detterman (Ed.), *Current Topics in Human Intelligence, Volume 5*. Norwood, NJ: Ablex.

Lakatos, I. (1970). Falsification and the methodology of scientific research programmes. In I. Lakatos & A. Musgrave (Eds.), *Criticism and the growth of knowledge* (pp. 91–92, 94–125, 127–138, 189–195). New York: Cambridge University Press.

Sternberg, R. J. (1977). *Intelligence, information processing, and analogical reasoning: The componential analysis of human abilities*. Hillsdale, NJ: Erlbaum.

Thurstone, L. L. (1938). Primary mental abilities. *Psychometric Monographs, 1*.

Vernon, P. E. (1950). *The structure of human abilities*. New York: John Wiley & Sons.

Wagner, R. K., & Oliver, W. L. (in press). How to get to Carnegie Hall: Implications of exceptional performance for understanding environmental influences on intelligence. In D. K. Detterman (Ed.), *Current Topics in Human Intelligence, Volume 5*. Norwood, NJ: Ablex.

Wechsler, D. (1944). *The measurement of adult intelligence*. Baltimore: The Williams & Wilkins Co.

Chapter 9

'g', Genes and Pedagogy: A Reply to Seven (Lamentable) Chapters

Chris Brand

Department of Psychology, University of Edinburgh

Assuming they are representative, the seven chapters that I have seen reveal a sorry state of affairs. Although they appear professional and half-competent, they are largely—not wholly, of course—neglectful of today's main developments in the study of general intelligence (*g*), pusillanimous about the significance of modern knowledge of *g*, and frankly devoid of any positive content of their own. Here I will first spell out these three main criticisms and then turn to specific matters such as: the unfairness of many current ('group') IQ tests to Asians; how to boost IQ, if desired; the unimportance of practice in accounting for human performance at cognitive tasks; and the existence of 'zonal' effects that require children of different IQs to be treated differently (i.e., educationally streamed).

NEGLECT

These chapters about intelligence largely ignore the major methodological advances, scholarly names and research findings of modern times. (a) The authors—with Pianta & O'Connor as honorable exceptions—eagerly attribute parent–child behavioral correlations (*r*'s) to the environmental influence of parents on children. They forget that the *different* genotypes (e.g., of siblings) will elicit, select, and create different microenvironments; and they forget that parent–child phenotypic *r*'s can easily reflect the influences of a third causal influence, the parental genotype. To ignore both the tide-turning review by Bell (1968) *and* the long standing logical analysis of *r*'s by J. S. Mill is to perpetuate the worst behaviorist, environmentalist biases of twentieth-century psychology. Most parents realize the importance of genetic factors once they see how their second child differs from the first. Why do tenured academics so often decline even to *consider* genetic and individual-centered explanations? (b) The work of Arthur Jensen, Ted Nettelbeck, Mike Anderson and Tom Bouchard is likewise largely ignored. The chapter authors decline to use the classic terminology of Spearman's *g* or of Cattell's distinction between fluid (*gf*) and crystallized (g_C) intelligence—preferring instead such novelties as "intelligence competence" and "social competence (with social status partialed out)." They ignore the substantial *r*'s between IQ and inspection time (IT) (speed of apprehension of the simplest information) even in nonretarded young adults—let alone the higher *r* of .70 that obtains between *g* and IT across the *full* population range. They prefer to sound off about going "beyond contextual social addresses" to forthcoming "detailed dissections" of omnipresent "interaction effects," about "the inaccuracy of bipolar [nature–nurture] debate" and about "the complexity of environmental systems" rather than admit modern evidence that *g* has a broad heritability of 70% and that, since fraternal twins do not differ dramatically in *g* when reared apart, individual differences in *g* are *not* substantially affected by G × E interaction effects (Bouchard, Lykken, McGue, Segal, & Tellegen, 1990; Brand, 1993; Pedersen, Plomin, Nesselroade, & McClearn, 1992). (c) Just as incredibly, the important adoption study by Schiff and others (Capron & Duyme, 1989; Schiff & Lewontin, 1986) is hardly mentioned, though this may be because of the embarrassment for environmentalists that the children adopted into French upper-middle-class homes received no more of an IQ boost in *gf* than Eysenck and Jensen had long predicted (Brand, 1987a).

PUSILLANIMITY

It is thus hardly surprising that the authors ignore Montie and Fagan's (1988) report of a large, 13.5 IQ-point intelligence difference between Black and

White three-year-olds *even though* the Black and White mothers had been matched for educational levels, medical histories, and marital status. The authors' neglect of the major themes in the study of intelligence today is so gross that it must be motivated—and the motivation is apparently that of fear. These authors respect conventional pieties that hereditarianism is "over-simplistic" and they burble that everything is a "complex, ongoing, multifaceted interaction effect" into which "more research is needed." They reach for such gobbledegook because admitting the main facts about *g* today would spoil their applications for state grants and mean they had to go to the Pioneer Fund. So, obviously, they repress America's best-established, most important, and environmentally unexplained group difference.

VACUITY

Presumably, the authors might have been braver if they had any very clear or promising ideas of their own. As it is, their writing abounds in "maybes," pious hopes, buzzwords, weak effects, waffle, and the use of countless references (often to unpublished or unexplained work) as a substitute for argument. The fact is that from none of these chapters could parents extract any reliable advice as to how to increase the chance that their children will be bright. What the authors won't admit is that both biological parents should themselves have highish IQs—with all the additional capacity for attentive, responsive, sensitive, and subtle parenting that a higher IQ involves, quite without psychologists' assistance.

ASIAN ACHIEVEMENTS

The achievements of Asians provide the chief exception to the general rule that the incoming data of today have made social environmentalists retreat beneath a smokescreen of verbiage from their now collapsed historical positions. Flynn, (Chapter Two, this volume) tantalizingly weaves his attribution of Asian success to "hard work" together with his claim that the worldwide secular IQ rise shows the unimportance of measured intelligence (since he takes the rise to have had no tangible consequence). With the help of his painstaking scholarship, determination, good humor and tricks with mirrors he seeks to entrance his readers into a happy valley in which the Black deficit in IQ and conventional achievement (for which he bravely admits to have no obvious environmental explanation—merely an as yet *non*-obvious one) could seem on the very verge of vanishing into the ether. Alas, the fifteen years that Flynn has labored to find the environmental 'Factor X' to explain the Black-White difference have yielded precisely nothing but this pleasing illusion, as any reader will know who can remember Flynn's original question. Instead, it cannot be long now before Flynn arrives at my own positon that many (group) IQ tests are simply unfair to Asians (and other conscientious folk) because they place a premium on (output) speed rather than on accuracy of performance, and that the worldwide rise in IQ-test scores reflects mainly the increasing impulsiveness, permissiveness, and casual tolerance of errors that are also manifest in the twentieth century's steadily rising rates of crime, sexual perversion, sexual disease, and family breakdown (despite enormous technical improvements that previous generations would have expected to yield utopian mastery of such problems). Flynn thinks my hypothesis of a changed speed (accuracy trade-off) to be falsified by the fact that IQ rises are sometimes seen on untimed tests. However, these gains are much lower (as I see Flynn beginning to admit); and Flynn neglects what I have explained to him (Brand, 1989, 1990), that tests like the Raven's Matrices, even if given untimed, still involve no correction for guessing, yet actually *require* guessing by their very structure. (They have items ascending in difficulty within each subscale, thus penalizing testees who tire themselves and waste scoring chances by scrupulously avoiding giving guesses to the harder items on earlier subscales). Notoriously, people of a more cautious and conservative outlook perform poorly on such tests (Brand, 1987b). (It is entirely reasonable that some IQ tests should penalize the painstaking, scrupulous testee having high personal standards and wishing to give defensible answers rather than "best guesses." The mass use of tests evolved precisely to penetrate the veneer of education and overachievement and to cut the hypereducated testee down to size). Flynn is a quick enough learner to have abandoned his erstwhile argument that the mere 1½ point IQ-score gain in Scotland over twenty years indicates, because of "regression of gain scores to the mean," a "true" IQ gain of an amazing 10 points, so I am sure he will soon come round to a unified theory along the above lines to account for *both* IQ rises *and* Asian overachievement. (None of this is to deny that *some* real *g* gain has occurred this century—mainly because of massively increased outbreeding thanks to the bicycle (Jones, 1993), the motor car, and the "university" encampments and rock concerts that, since 1920, have all come to bring young people together en masse across once largely unbridged distances). In its combination of energy with grievous logical error (Brand, 1990), Flynn's scholarly program is a true product of the permissive society that values febrile "creativity" more than the sustained productivity that finds favour with Asians.

HEADSTART PROGRAMS

Headstart provides another topic on which the present authors have to exercise massive repression in order to protect their flaky academic egos. (In this matter, Flynn is *not* guilty—indeed, his chapter provides some rejoinder to Wagner's (Chapter Seven, this volume) quack claims based on extreme manipulations that cannot be mimicked in education.) Not only is Spitz's (1986) work unmentioned, or that of Clarke and Clarke (1989); but these enthusiasts for conventional, "social," IQ-raising efforts fail to recognize that the expert-selected adoptive family fulfils the wildest dreams of most Headstarters and yet, to this day, does very little for Black or other low-IQ children (Lynn, 1994; Weinberg, Scarr, & Waldman, 1992). At present, the only proven way of boosting IQ is by drug-free breast feeding: this yields a 10-IQ-point boost even when mothers are matched for social class and the initial intention to breast feed (Lucas, Morley, Cole, Lister, & Leeson-Payne, 1992). However, breast feeding is not popular with lower-IQ mothers (or with their Oedipally jealous husbands or partners); and researching it will not prove very job-creative for psychologists who don't like IQ testing, so this finding, too, must be repressed.

If practice did indeed make perfect, there would be a case for continuing the so far unavailing search for a workable program of the Headstart type. Yet, once more, the review of the literature by the present authors is strangely selective. There is overwhelming evidence that important human skills are loaded for *g* and not much else (Hunter & Hunter, 1984); and the uncited Arthur Jensen (1991, 1993) finds almost a basic "law" (1993, p. 178) whereby learning gains are higher in the more intelligent learners (as Wagner & Oliver come close to admitting). Again, Ackerman's (1987) paper—so central to the argument of Chapter 7—records that there *are*, in recent studies, "correlations between performance measures on skill learning tasks and standard intellectual and cognitive ability measures" (p. 3), quite contrary to Wagner and Oliver's main line of argument.

Of course the *broad heritability* for IQ is substantially higher than the narrow heritability (the "breeding-true" heritability)—as is found for other biological variations. [As Cyril Burt realized, the *narrow* heritability for IQ which he probably overestimated a little at .52 (Burt & Howard, 1971, p. 15, Table V) could not yield a very satisfactory caste system unless accompanied by big environmental differences, as in postimperial India.] Moreover, the importance of *unshared, within-family variance* is much more noticeable on *non-g* personality traits than on *g* itself: Hoffman (1991) adduced sound enough "theoretical" reasons for expecting birth order to be an important variable (as Adler supposed–see Brand, 1994), but the truth is that birth order has virtually *no* effect on *g* itself (once family size is controlled) (see Bouchard, 1993).

Yet pessimism about human improvement is not the correct response to a proper modern realization of g's heritability and importance. Just as psychological phenomena (such as *r*'s between variables) quite often look different in different *g* ranges (Brand, 1994; Brand, Egan & Deary, 1993;Detterman & Daniel, 1989), so *g*-level is itself a vital mediator of response to particular types of instruction (Carroll, 1993; George, 1990; Nemko, 1988; Swing, 1994). Educational streaming has been largely abandoned in U.S. and British schools out of egalitarian fervor; this has done nothing, however, to equalize children either in *g* or in educational achievement, and it has lowered educational attainment as compared to the streamed educational systems of France, Germany and Northern Ireland. It is now time to allow all children and parents the *choice* of different levels of instruction for children and to break open Western society's last slave camps—for children who are free to make choices at home are, in state schools, dragooned through ineffective, similitarian schooling for most of the hours of the compulsory school day. Perhaps this was what Pianta and O'Connor (Chapter 4, this volume) meant when mumbling about the possibility of "multi-level, multi-influence interventions." However, like other authors of these chapters, they were unwilling to spell out, think through, or champion even the most promising ideas. Such authors wilt before today's politically correct educational establishment of joyless intellectual deadbeats who spurn the historical individualism of the English-speaking people. How the dismal prophets of modern piety listen to the siren song of Ibsen's *Provost*—"We want all men to be equal . . . The surest way to destroy a man is to turn him into an individual. Very few men can fight the world alone (Ibsen, 1866/1986, p. 93)." Instead, it is time now for psychologists to admit what is known and to agree with Havender (1987):

> schools wishing to improve the effectiveness of education for all children will have to adjust their curricula to their students, not continue to expect their students to fit into a uniform, Procrustean educational bed in the mistaken belief that they are limitlessly malleable (p. 341).

The present chapters advance no cause of significance—scientific or educational; but in their defensiveness and denial they leave little *g* unscathed and ready to guide all those educators who genuinely want to do the best for each individual child. Even B. F. Skinner (1984, p. 951) came round to decrying "the educational phalanx" and urging that each student "move at his or her own pace." It is sad to find putative experts on intelligence wallowing in unreconstructed 1960s academic psychobabble. It is worse still to realize that they are protected from exposure by applications for further State "research" hand-outs.

REFERENCES

Ackerman, P. L. (1987). Individual differences in skill learning: An integration of psychometric and information-processing perspectives. *Psychological Bulletin, 102*, 3–27.

Bell, R. Q. (1968). A reinterpretation of the direction of effects in studies of socialization. *Psychological Review 75, 2*, 81–95.

Bouchard, T. J., Jr., Lykken, D. T., McGue, M., Segal, Nancy L., & Tellegen, A. (1990). Sources of human psychological differences: The Minnesota study of twins reared apart. *Science, 250*, 223–228.

Bouchard, T. J., Jr. (1993). Genetic and environmental influences on adult personality: Evaluating the evidence. In J. Hettema & I. J. Deary (Eds.), *Foundations of personality* (pp. 15–44). Dordrecht: Kluwer.

Brand, C. R. (1987a). A touch of (social) class for psychology. *Nature, 325*, 767–768.

Brand, C. R. (1987b). British IQ: Keeping up with the times. *Nature, 328*, 6133, 761.

Brand, C. R. (1989). Has there been a "massive" rise in IQ levels in the West? Evidence from Scottish children. *Irish Journal of Psychology, 10*, 3, 388–394.

Brand, C. R. (1990). A "gross" underestimate of a "massive" IQ rise? A rejoinder to Flynn. *Irish Journal of Psychology, 11*, 1, 52–56.

Brand, C. R. (1993). Cognitive abilities: current theoretical issues. In T. J. Bouchard, Jr., & Propping, P. *Twins as a Tool of Behavioral Genetics* (pp. 17–32). Chichester: Wiley.

Brand, C. R. (1994). Dimensionalizing personality: the 'Big 5', the 'Gigantic 3' and the 'Comprehensive 6'. *Psychologica Belgica, 34*(4), 257–275.

Brand, C. R. (in press). Personal interest and social interest: Alfred Adler's dimensions of personality. *Behavior Research and Therapy.*

Brand, C. R., Egan, V., & Deary, I. J. (1993). Personality and general intelligence: aboriginal realities or 'language games'? In G. L. Van Heck, P. Boraiuto, I. J. Deary, & W. Nowack (Eds.), *Personality psychology in Europe 4.* (pp. 203–228). Tilburg: Tilburg University Press.

Burt, C., & Howard, M. (1971). Quantitative genetics in psychology. *British Journal of Mathematical and Statistical Psychology 24*(1), 1–21.

Capron, C., & Duyme, M. (1989). Assessment of effects of socioeconomic status on IQ in a full cross-fostering study. *Nature, 340*, 552–554.

Carroll, J. B. (1993). *Human cognitive abilities.* Cambridge: Cambridge University Press.

Clarke, A. M., & Clarke, A. D. B. (1989). The later cognitive effects of early intervention. *Intelligence, 13*, 289–297.

Detterman, D. K., & Daniel, M. H. (1989). Correlations of mental tests with each other and with cognitive variables are highest for low-IQ groups. *Intelligence, 13*, 349–359.

George, D. (1990). The challenge of the able child. *Cambridge Journal of Education, 20*(2) 175–182.

Havender, W. (1987). Educational and social implications [of IQ]. In S. Modgil & C. Modgil (Eds.), *Arthur Jensen: Consensus and controversy* (pp. 339–351). Brighton: Falmer.

Hoffman, L. (1991). The influence of the family environment on personality. *Psychological Bulletin, 110*, 187–203.

Hunter, J. E., & Hunter, R. F. (1984). Validity and utility of alternative predictors of job performance. *Psychological Bulletin, 96*(1), 72–98.

Ibsen, H. (1866/1986). *Brand.* London: Methuen.

Jensen, A. R. (1991). Spearman's *g* and the problem of educational equality. *Oxford Review of Education, 17*, 169–187.

Jensen, A. R. (1993). Psychometric *g* and achievement. In B. R. Gifford (Ed.), *Testing and public policy* (pp. 117–227). Boston: Klewer.

Jones, S. (1993). *The language of the genes: Biology, history and the evolutionary future.* London: HarperCollins.

Lucas, A., Morley, R., Cole, T., Lister, G., & Leeson-Payne, C. (1992). Breast milk and subsequent intelligence quotient in children born pre-term. *Lancet, 339*(8788), 261–264.

Lynn, R. (1994). Some reinterpretations of the Minnesota transracial adoption study. *Intelligence, 19*(1), 21–28.

Montie, J. E., & Fagan, J. F., III. (1988). Racial differences in IQ: Item analysis of the Stanford–Binet at three years. *Intelligence, 12*, 315–332.

Nemko, M. (1988). The report card. *Montclarion, 23*, vi.

Pedersen, N. L., Plomin, R., Nesselroade, J. R., & McClearn, G. E. (1992). A quantitative genetic analysis of cognitive abilities during the second half of the life span. *Psychological Science, 3*(6) 346–352.

Schiff, M., & Lewontin, R. (1986). *Education and class: The irrelevance of IQ genetic studies*. Oxford: Clarendon.

Skinner, B. F. (1984). The shame of American education. *American Psychologist, 39*, 947–954.

Spitz, H. (1986). *The raising of intelligence*. Hillsdale, NJ: Erlbaum.

Storfer, M. (1990). *Intelligence and Giftedness*. San Francisco: Jossey-Bass.

Swing, Elizabeth (1994). Review of F. J. Monks, M. Katzk, & H. W. Van Boxtel (Eds.), 1992. *The education of the gifted in Europe: Theoretical and research issues*, Amsterdam: Swets and Zeitlinger. *Corporative Education Review, 38*, 2, 282–284.

Weinberg, R. A., Scarr, S., & Waldman, I. D. (1992). The Minnesota transracial adoption study: A follow up of IQ test performance at adolescence. *Intelligence, 16*, 117–135.

Chapter 10

Commentary on Chapters 1–7

Ann M. Clarke
Alan D. B. Clarke

University of Hull, United Kingdom

We were intrigued by the prospect of reviewing these seven chapters, believing that, as the editors indicated, there would be "something for everyone to disagree with." On the whole, however, most authors seem to have presented topics from a safe perspective, and there is thus rather little to rebut. Nevertheless, we were surprised at Keating's (Chapter 3) presentation of "nature" and "nurture" as if there still existed in the minds of many research psychologists an important dichotomy between them. Surely we may all agree with him that both are fully integrated in ontogenesis, while being interested in the efforts of those who choose to discover which may be the more important in differing contexts. Such approaches may avoid our being misled into believing "self-evident truths."

So far as the majority of families are concerned, behavior genetic research has shown that, for a wide variety of traits that include measures of intelligence, specific cognitive abilities, personality, and psychopathology in North American and European populations, heritabilities lie between .40 and .70

(Scarr, 1992). Of the remaining reliable variance, there is more variation within families than between families. The majority provide sufficiently supportive environments for their children to develop well cognitively. We need constantly to remind ourselves that, in the long term, shared environmental influences in the family have no lasting impact upon IQ differences, the correlation between older adoptive siblings being zero (Plomin & Daniels, 1987; Scarr & Weinberg, 1978).

Adoption studies may provide evidence on two different questions: (a) the correlation within families among biologically unrelated, but socially very close individuals, and (b) the levels achieved in terms of IQ, scholastic achievement, and adjustment of adopted individuals compared with their biological siblings remaining in disadvantaged circumstances. The latter (Dumaret, 1985) provides good evidence for environmental influences, as do a series of studies by Tizard and colleagues (e.g., Hodges and Tizard, 1989) contrasting the later development of a cohort of children from an institution, some of whom remained there, some who have returned to their disadvantaged circumstances, and some of whom were adopted after age 2. The findings have been supplemented by case studies of children rescued relatively late from terrible circumstances (Clarke & Clarke, 1976; Skuse, 1984).

Ever since the 1960s, planned educational and social intervention for children raised in disadvantaged conditions have been studied by many behavioral scientists. We have summarized a great deal of evidence concerning the later cognitive effects of early intervention (Clarke & Clarke, 1989) emphasizing that, as time passes after program termination, the earlier cognitive increments follow the law of diminishing returns, unless the intervention sets off a chain of ongoing positive consequences. These may be as diverse as factors rendering the child more attractive to a member of the family group, resulting in more social transactions, or higher teacher expectations leading to a smaller likelihood of relegation to a special class.

Several authors in this volume (Keating; Pianta & O'Connor; Wachs, Wagner & Oliver;) note, with approval, Bronfenbrenner's seminal arguments about the importance for development of the social context, although perhaps they put too little emphasis on his insistence on "thoughout the life course" (Bronfenbrenner, 1989, p. 186). We agree, and indeed our statements across the years concerning intervention can perhaps best be interpreted in this light. We have, of course, never denied the short-term gains in IQ and achievement to children who had followed preschool intervention programs. Nor would we seek to detract from the pleasure and pride that presumably occur in this connection. However, we continue to hold that the constant pressure from the micro-, meso-, and macroenvironments on the severely disadvantaged child over the life span is likely, actuarially, to counteract the positive effects of such programs unless they maintain their hold over children during the school years (as in part of the Abecedarian cohort, Ramey, 1992) or initiate a sequence of positive ongoing events as in the well-publicized Perry Preschool Program (Berrueta-Clement, Schweinhart, Barnett, Epstein, & Weikart, 1984; Schweinhart & Weikart, 1980, 1981). May we thus suggest that future results be quoted not only as significant differences between experimental and control groups (hopefully favoring the former), but also their actual status in comparison with national norms.

Ramey & Blair (Chapter 5) are, of course, aware of "washout" effects after termination of intervention programs, and they stress the necessity of "bridging" or "carrier" mechanisms to reinforce progress. Increases in intellectual functioning, they write, can be maintained "through continued mentoring across a variety of culture-specific domains of knowledge," a premise with which we agree. We agree, too, with these authors' approval of Lazar & Darlington's (1982) analysis of the outcome of very high quality early intervention programs, with their significant reduction in grade retention and special class placement. No doubt, these owed their effects to some degree of bridging processes, although the extent of the intervention outcome should not be overemphasized. Ramey (1982) himself, commenting on this study, indicated that:

> the mean IQ performance at follow-up . . . is approximately one standard deviation below the national average, for both program and control children. Clearly, then, this represents a group of children who are likely to experience major hardships in an increasingly technological and sophisticated culture. That these results obtained in spite of efforts of some of our leading social scientists and educators testifies to the difficult and complex set of conditions associated with lower socio-economic status in this country . . . we are unlikely to witness full realization of human potential through limited educational experiments. (p. 149)

Large-scale studies of disadvantaged communities indicate that there may be "spontaneous" advances by some individuals reared in very adverse circumstances (Kolvin, Miller, Scott, Gatzanis, & Fleeting, 1990; Werner, 1989). We do not know for sure what the precise mechanisms may be, but there are some clues summarized by Clarke & Clarke (1992). In a greatly abbreviated form, protective factors include constitutional predispositions towards alertness, attractiveness, sociability, problem-solving ability, and purposeful planning, leading to an internal locus of control. The social contexts include some network of affectionate support, positive schooling, and prosocial peer groups, all of which may be seen as being evoked by the developing individual.

We are in substantial agreement with Borkowski & Dukewich (Chapter 1) in emphasizing attachment as one ingredient in later cognitive outcome. We would also suggest the beneficial effects of attachment arising later than infancy, including, as Rutter (1989) has pointed out, happy marriages following earlier social deprivation.

We found Flynn's chapter (7) to be refreshingly controversial, especially his provocative argument contrasting massive international secular trends in IQ with real gains in intelligence. Moreover, he seeks to demolish attempts to identify environmental factors (over and above test sophistication and test techniques) involved in these increases. These include improved nutrition, eradication of childhood diseases, and improvements in preschool and home environments. None of these, he suggests, can explain the demonstratable cross-national gains. As usual, Flynn states his position boldly and clearly; he believes that the portion of IQ gains over time that represents real intellectual gain is very small indeed, and that, paradoxically, the exciting thing to explain is the huge nonintelligence (IQ test) increment. His is a chapter where everyone will find something to agree with—and something to disagree with, if only we had the evidence!

The interlinking chain of biological, social, and personal factors (including motivation, need for achievement, and locus of control) that influence individual development is only now beginning to be unravelled. In the meanwhile, let us not be tempted to abandon IQ scores with their long-established correlates!

All seven chapters discuss at least some of the mechanisms of environmental influences on cognitive development. The theoretical contributions of Pianta & O'Connor (Chapter 4) and Wachs (Chapter 6) both emphasize the complexity of environmental effects. The former draw attention to systems models that incorporate biological influences, arguing that many studies applying "ideas of environmental specificity mis-specify the complex nature of development" (Pianta & O'Connor). Research is needed that takes account of developmental niches (genes and environment) for IQ. As an example, they state that temperament may influence environmental experiences, which are, in turn, related to cognitive development. In this way, it is surprising that the authors fail to note the long-term work of Thomas & Chess (1980), who from the 1960s on have emphasized the bidirectional role of temperament in environmental interactions. Wachs, too, draws attention to "environmental systems" as opposed to "the environment" (this volume). These co-vary with genetic factors, nutritional status, morbidity, and exposure to toxins. Few would question these points.

Wagner and Oliver's chapter (7) is another provocative contribution. Their thesis is that exceptional performance in any skill results from a decade or more of intensive training, and depends upon an unusual combination of resources, effort, and motivation. They argue that key principles derived from the study of extraordinary proficiency might have their counterparts in the case of environmental influences on intelligence. Short-term practice on both simple and complex cognitive tasks results in improved performance. Further, they claim controversially that "the long-term practice provided by schooling appears to influence the development of subsequent intelligence" (Wagner & Oliver, this volume). The authors acknowledge that not everyone can achieve skilled performance in any domain, nor that individuals profit from training to the same degree—merely that skilled performance and intelligence each represent "the culmination of the actions, and interactions, of genetic and environmental influences over a prolonged developmental period," an exceptionally modest conclusion to pages of block-busting arguments, the implications of which are that experts are made, not born.

Wagner & Oliver should perhaps consider the burgeoning literature on the extraordinary specific abilities of people of low IQ—the so-called idiots savants. These people flower early (without a decade or more of intensive training), and, as O'Connor & Hermelin (1988) put it, "One of the very notable features of the occurrence of talents in the idiot savant is that they frequently emerge unbidden, usually between ages 5 and 8 years, often on the basis of no detectable genetic influence and *in the complete absence of training* [italics added]" (pp. 393–394). The issue of improvement with practice over time was specifically addressed in connection with two ten-year-old calendrical calculators, having IQs of around 90 (O'Connor & Hermelin, 1992). It was concluded that these children had already inferred rules about calendrical structure, that their performance could not be accounted for by practice alone, but also by the use of (quite exceptional) cognitive strategies. How are such findings to be accommodated with the thesis advanced in Chapter 7, or aren't they? And can training, even at the hands of outstanding teachers, really enhance intelligence, as opposed to a more narrowly defined skill? Perhaps Wagner & Oliver's thesis could be further tested in empirical studies leading to applied outcomes—or should we leave these immense problems to politicians?

REWFERENCES

Berrueta-Clement, J. R., Schweinhart, L. J., Barnett, W. S., Epstein, A. S., Weikart, D. P. (1984). *Changed lives: The effects of the Perry Preschool Program on youths through age 19*. (Monographs of the High/Scope Educational Research Foundation). Ypsilanti, MI: High/Scope Press.

Bronfenbrenner, U. (1989). Ecological systems theory. *Annals of Child Development, 6*, 187–249.

Clarke, A. M. & Clarke, A. D. B. (1976). (Eds.). *Early experience: Myth and evidence.* London: Open Books.

Clarke, A. M. & Clarke, A. D. B. (1989). The later cognitive effects of early intervention. *Intelligence, 13*, 289–297.

Clarke, A. M. & Clarke, A. D. B. (1992). How modifiable is the human life path? *International Review of Research in Mental Retardation, 18*, 137–157.

Dumaret, A. (1985). IQ, scholastic performance, and behavior of sibs raised in contrasting environments. *Journal of Child Psychology and Psychiatry, 26*, 553–580.

Hodges, J. & Tizard, B. (1989). IQ and behavioral adjustment of ex-institutional adolescents. *Journal of Child Psychology and Psychiatry, 30*, 53–75.

Kolvin, I., Miller, F. J. W., Scott, D. McL., Gatzanis, S. R. M., & Fleeting, M. (1990).

Continuities of deprivation? The Newcastle Thousand Families Study. Aldershot, UK: Gower House.

Lazar, I., & Darlington, R. (1982). Lasting effects of early education: A report from the Consortium of Longitudinal Studies. *Monographs of the Society for Research on Child Development, 47*, 2–3, Serial No. 195.

O'Connor, N. & Hermelin, B. (1988). Low intelligence and special abilities. *Journal of Child Psychology and Psychiatry, 29*, 391–406.

O'Connor, N. & Hermelin, B. (1992). Do young calendrical calculators improve with age? *Journal of Child Psychology and Psychiatry, 33*, 907–912.

Plomin, R. & Daniels, D. (1987). Why are children in the same family so different from one another? *Brain and Behavioral Sciences, 10*, 1–60.

Ramey, C. (1982). Commentary. In Lazar, I., & Darlington, R. *Monographs of the Society for Research on Child Development, 47*, 142–151. (Serial No. 195).

Ramey, C. (1992). High-risk children and IQ: Altering intergenerational patterns. *Intelligence, 16*, 239–256.

Rutter, M. (1989). Pathways from childhood to adult life. *Journal of Child Psychology and Psychiatry, 30*, 23–51.

Scarr, S. (1992). Developmental theories for the 1990s: Development and individual differences. *Child Development, 63*, 1–19.

Scarr, S. & Weinberg, (1978). The influence of "family background" on intellectual attainment. *American Sociological Review, 43*, 674–692.

Schweinhart, L. J. & Weikart, D. P. (1980). *Young children grow up: The effects of the Perry Preschool Program on youths through age 15*. (Monographs of the High/Scope Educational Research Foundation). Ypsilanti, MI: High/Scope Press.

Schweinhart, L. J. & Weikart, D. P. (1981). Perry Preschool effects nine years later: What do they mean? In M. J. Begab, H. C. Haywood, & H. Garber (Eds.), *Psychosocial influences on retarded development* (pp. 113–125) Baltimore: University Park Press.

Skuse, D. (1984). Extreme deprivation in early childhood, II: Theoretical issues and a comparative review. *Journal of Child Psychology and Psychiatry, 25*, 543–572.

Thomas, A., & Chess, S. (1980). *The dynamics of psychological development*. New York: Brunner/Mazel.

Werner, E. E. (1989). High-risk children in young adulthood: A longitudinal study from birth to 32 years. *American Journal of Orthopsychiatry, 59*, 72–81.

Chapter 11

Misrepresentations and Distortions in Second-Hand Accounts of Research

Howard L. Garber

Milwaukee Center for Independence

James Hodge

Madison, Wisconsin

It is understandable that most students of psychology can not fully gain their knowledge and understanding of the field by primarily reading original sources of information. They often have to rely instead on interpretations of classic theory or popularly cited experiments in textbooks, professional literature reviews in books, or citations in journal articles. Unfortunately, these second hand accounts often feature among the facts considerable fabrication and distortion. For example, Blumenthal (1975) and Bringmann, Balance, and Evans (1975) cautioned that most textbook portrayals of the many accomplishments of

Wundt, who is generally recognized as experimental psychology's great patron, are far from historically accurate. Todd and Morris (1992) noted that Watson and Skinner both expressed frustration over the continued misrepresentations of their respective theories of behaviorism in the scientific as well as the popular literature, and lamented the waste of time and effort required to correct these distortions. Arthur Jensen, as a result of his now classic review of behavioral genetic research, spent almost two decades refuting contentions that he believed experience had no influence on intellectual development, and that he attributed all variability in intelligence to genetic differences (Jensen, 1969).

Four critical examinations of descriptions of Watson's widely cited emotional conditioning experiment (Cornwell & Hobbs, 1976; Harris, 1979; Prytula, Oster, & Davis, 1977; Samelson, 1974) suggested that the details of this study are not only badly known, but have been obscured and distorted in second- and third-hand accounts. A reanalysis of the literature 10 years later (Paul & Blumenthal, 1989) indicated that there had been little effort to correct these inaccuracies. Similar errors have been found in descriptions of other classic or widely cited studies such as Breuer's treatment of Anna O. (Ellenberger, 1972) and Pavlov's conditioning experiments (Goodwin, 1991).

In the retelling of research, the relatively minor errors in detail which appear in most second-hand accounts are of little concern because they do not interfere with the honest efforts to evaluate and interpret theoretical positions or findings. If, however, the reliance on second-hand accounts is predominant and those secondary sources include major misrepresentations and distortions, as many do, they encourage a sense of familiarity with the purpose, procedure and outcome of an investigation which is false. Inaccuracies, encountered repeatedly, take on a life of their own, because second-hand sources of information are rarely verified and are assumed correct even though they differ significantly from original accounts. This, of course, leads to inaccurate interpretations and a lack of appreciation of the real significance of theory and research findings and ultimately to the development and perpetuation of myths.

Myth making associated with a particular investigation need not be viewed as a process whereby a conscious attempt is made to deceive, but may reflect the operation of rather passive influences. A number of factors can generate distortion and misrepresentation in second-hand accounts of research. Some errors can be attributed to inconsistencies in original reports. Others can be attributed to poor scholarship. Many errors simply reflect a drift away from the original due to retelling or simplification for the convenience of explanation and pedagogy. All second-hand accounts seem to leave something out of their descriptions, and the emphasis or importance placed on others is simply a matter of selective judgment. Ultimately, however, variations in retelling and in interpretation of research are presumed an honest representation. There are, however, more active or dynamic influences which may account for distortion or misrepresentation in second-hand accounts of research by the action of reviewers. These occur when reviewers choose to include or not include in descriptions of research, material selected according to one's own theoretical bias or a need to establish or refute evidence for a particular position or interpretation, which results in active revision of the original purpose of the investigation (Harris, 1979). Such descriptions that misrepresent or distort crucial aspects of theories or investigations significantly contribute to myths perpetuated historically and cannot be overlooked.

On a broader perspective, the general trend of thought and theory that dominates an area of study or research during specific periods can create general biases in how it is perceived, described, and interpreted, regardless of the original intent (Blumenthal, 1975). Preconceived ideas about an investigation often preclude objective consideration of the facts. The need to build a sense of cohesiveness and continuity in a particular area of study may result in the minimization of differences, overgeneralization of similarities and a disregard for discrepancies in order to provide simplified and presumedly unambiguous, either/or explanations and examples (Cornwell & Hobbs, 1976; Harris, 1979). The intent of an investigation does not necessarily have to be consistent with the zeitgeist of the period within which it is conducted or within which it is interpreted.

Second-hand accounts and critiques of our own research, the Milwaukee Project, are almost universally inaccurate to some degree. Garber's (1988) final report provides a full account of the details and the results of this study of the effects of being raised by a retarded mother on the rate of children's intellectual development. Jensen (1989) provided an in depth critique of this volume, and four additional critiques appeared as a special book review section of the American Journal on Mental Retardation (Butterfield & Berkson, 1991). These reviews were responded to by Garber and Hodge (1989) and Garber, Hodge, Rynders, Dever, and Velu (1991).

This chapter is not concerned with the details of the longitudinal investigation or the methodology questions raised in these critiques or critiques of the project that appeared before the final report was published (viz., Page, 1972, 1986; Page & Grandon, 1981). The topic at hand is the manner in which the factors we have discussed may have influenced the almost universal distortion and misrepresentation of the theoretical basis, purpose, and general findings of the Milwaukee Project in these critiques as well as in other second-hand accounts of the research. This chapter seeks to clarify confounding issues but cannot resolve the process of interpretation and explanation that results in confusion between research problems which allows or encourages distortion, inaccurate representations or unwarranted interpretations. The result, quite unfortunately, is the extent that this problem has contributed to the slowing of resolve or the development of more clarity on matters of intellectual development for children at risk. In our presentation we describe and give examples of how this process has occurred.

INCONSISTENCIES IN THE ORIGINAL REPORTS

The Milwaukee Project longitudinal investigation came under considerable early criticism (Page, 1972, 1986; Page & Grandon, 1981) because Rick Heber, the principal investigator and director of the project, was slow to provide adequate and timely detailed reports of results. This criticism was, for the most part, legitimate. It could be argued that inaccuracies in second-hand accounts are attributable to the fact that original reports of the research were limited to conference proceedings, book chapters, progress reports, and dissertations that contained insufficient detail and/or summaries of major findings. This does not, however, explain the misrepresentation and distortion of the general purpose or the research outcomes in critiques and second-hand accounts of the research.

All original accounts of the investigation, regardless of how brief, consistently included descriptions of the general findings of the survey study on which the longitudinal research was based. These descriptions also included statements that the purpose of the intervention was to prevent declines in IQ across age for normally functioning infants born to mentally retarded mothers. Garber's (1988) final report of the project provides a full account of the details and results of the investigation. Any lack of detail or inconsistency in early reports clearly does not explain misrepresentations and distortions found in reviews of this report (Butterfield & Berkson, 1991; Jensen, 1989) or in articles written after this report was published (viz., Flynn, this volume; Jackson, 1993; Locurto, 1988, 1991). These distortions appear to be mainly the product of preconceived ideas about the research or personal bias in interpretation.

Authors' Personal Bias In Reassessing Environmental Influences on Intelligence

Flynn's chapter, in this volume, provides a recent review of literature that seems to be an example of a review made in an attempt to support a theoretical point and actually contains considerable distortion, misrepresentation, and omissions. We find this discussion particularly troubling because our own research, the Milwaukee Project (Garber, 1988; Garber et al., 1991) is so badly misrepresented.

At a general level Flynn argues that the use of obsolete norms has complicated the assessment and interpretation of the impact of environmental factors on the interrelated traits of intelligence and achievement. Flynn's data indicate that if subjects are administered a test with obsolete norms their scores will be inflated in the sense that they will appear superior to their contemporaries when actually they are not. His contribution to the study of intelligence has been the development of formulae that facilitate adjustment of scores from different tests to a common standard (Flynn, 1984). Flynn (1994, this volume) attempts to demonstrate that scores from different tests of intelligence, as well as scores from the same test standardized at different times, are not directly comparable.

Contrary to Flynn, we believe that the major complication in exercises of comparability are not a result of obsolete norms or generational differences, but arise from the fact that all tests of intelligence, regardless of when they were normed, are based on different content, and different populations to establish norms for each test. Scores from different tests are rarely directly comparable. Furthermore, most studies of environmental influence on intelligence do not compare scores across generations, or even scores of individuals in study groups with the mean for their contemporaries in the general population. Most of these investigations entail comparisons of pre- and posttreatment assessments of the same individuals, or comparisons of scores for treated and untreated groups. The most important consideration in such cases is not whether or not the norms for the tests are current, but whether or not scores for all assessments are directly comparable, regardless of when norms were established for the tests.

An example of a misrepresentation occurs in Flynn's discussion of Vernon's (1982) review of studies of academic and occupational status differences between Chinese American and white American populations. Flynn begins by discussing Vernon's (1982) study of abilities and achievements of Chinese Americans, which was based heavily on previously unpublished data provided by Arthur Jensen. Unfortunately, his attempt here to provide evidence for his argument includes several examples of rather casual scholarship and a resultant misrepresentation. He makes a statement to the effect that Jensen believes Asian Americans may do well simply because they are smarter. Flynn not only does not reference Jensen directly but instead cites a general discussion of the topic of Asian American achievement by a science writer in *Time*, a news magazine (Brand, 1987).

Flynn badly distorts Vernon's findings and misrepresents his conclusions, which are actually quite similar to his own. Vernon concluded that the lack of obvious environmental differences between the two populations suggested that genetic differences may play an important role in differences in mean intelligence levels of the two populations. He cautioned, however, that there may be subtle differences in motivation and achievement values in these populations, which could account for differences in occupational success. Flynn's assertion that the differences in intelligence identified for Chinese American and white American populations in Vernon's review are the result of scores based on obsolete norms is incorrect. A majority of the studies that Vernon reviewed were comparisons of the performance of different ethnic groups on the same tests.

Flynn's discussion of both the Skodak and Skeels Adoption Study and our own research, the Milwaukee Project, does not include a single reference to an original source of information. He does reference the review of Garber's (1988) final report by Jensen (1989), but there is no evidence that he has read Garber's report or the response to Jensen's review by Garber and Hodge (1989), which appeared in the same journal issue. These errors of omission are uncharacteristic of Flynn's writing. However, such errors in any second-hand account of research should alert readers to the possibility of bias.

Flynn (1994, this volume) correctly observed that the mother–child IQ gap in the Skodak and Skeels Adoption Study had been overestimated because scores of the children were based on obsolete norms. The children and the biological mothers were tested on the 1916 Stanford–Binet; the children in 1946 and their mothers approximately 14 years prior to this. The scores of the children were therefore based on norms that were about 14 years obsolete in comparison to the norms used for scores for their mothers. Vernon's (1982) review, however, did not involve generational differences, or the comparison of scores for Chinese Americans with the mean of their contemporaries in the general population as implied by Flynn. Vernon compared scores of Chinese Americans with white Americans taking the same tests at approximately the same time. Unfortunately, in Flynn's (1984) analysis, only the IQs of Chinese Americans, not white Americans, were adjusted. The scores of white Americans would have been equally inflated because they took the same tests. Obsolete norms in no way complicate the interpretations made by Vernon of possible differences between these two groups. As discussed earlier, the major concern in such comparisons is not whether or not the norms are current, but whether or not the scores for all assessments are comparable.

Flynn (1994, this volume) makes a similar error in his descriptions of the Milwaukee Project. The major complication in the Milwaukee investigation was the comparability of tests on the same individuals across time because a number of different instruments were used to make assessments across the course of the study. We used Flynn's formulae to adjust scores from the different tests used in the Milwaukee Project so performance could be compared across tests (Garber & Hodge, 1989). The hypothesis tested in the Milwaukee investigation concerned declines in DQ and IQ across age, and the difference in the rate of intellectual development for treated and untreated children. The primary interest was not in making comparisons of the treated children's mean IQ with any population norm. Flynn (1984) does not even mention the control group in his analysis or discussion. Obsolete norms in no way complicated interpretation of findings of this investigation once scores from these tests were made comparable (Garber & Hodge, 1989).

Some readers may view this as a simple difference in interpretation. We believe, although it may be unintentional, that it represents a significant distortion of the original intent of these investigations and the hypothesis being tested. If the author of a second-hand account of research wishes to offer a different interpretation, care should be taken to accurately describe the hypothesis being tested and provide readers with accurate information needed to judge for themselves which interpretation has the most support.

Flynn's discussion of the Milwaukee Project in this volume contains two additional misrepresentations of the investigation that are more serious. He states that the intervention was (a) an attempt to give black children from the ghetto the advantages of an upper-class home, and that (b) it was believed by Heber and Garber that the mean IQ of these children had been raised to over 20 points above the average score of contemporary whites. These distortions of the purpose and outcome of the study have become part of the myth associated with the investigation that has been perpetuated through second-hand accounts across the years.

Attempts to correct misrepresentations of the Milwaukee Project have been frustrating. For example, Locurto (1988, 1991a) failed to recognize critical distinctions between the purpose of the Milwaukee Project and research efforts designed to raise the IQ and improve the academic performance of children living in poverty. Unfortunately, when we attempted to correct the distortions in his accounts of this research (Garber & Hodge, 1991), he not only refused to acknowledge his errors, but misrepresented several points we made in a follow-up response (Locurto, 1991b).

THE ZEITGEIST OF THE 1960S

Much of the distortion and the misrepresentation in second-hand accounts of the Milwaukee Project might be attributable to attempts to build a sense of cohesiveness and continuity in the area of early intervention research. Zigler and Hodapp (1986) discussed the development of the Zeitgeist or the general trend of thought and theory that dominated the study of intelligence, and particularly the study of mental retardation during the 60s and 70s. There was an almost universal attempt to explain the etiology of all intellectual retardation without known organic causes in terms of environmental factors.

There was also great social pressure to improve the relatively poor academic achievement of children living in the most economically disadvantaged areas of our society. The failure of these children to achieve was generally attributed to the fact that when they entered school their average intellectual level was significantly lower than the average of the general population. Researchers failed to identify any significant differences during infancy, but found progressively larger IQ differences at later ages. This cross-sectional finding was interpreted as a decline in IQ with increasing age. The phenomenon was accepted as a generalized characteristic of all intellectual retardation without known organic causes and, moreover, was almost uncritically attributed to a depression of the rate of intellectual development by the general conditions of poverty or social deprivation (Heber & Dever, 1970; Heber, Dever, & Conry, 1968).

This position, which served as the theoretical basis of much of the social policy and early intervention research conducted during this period, overlooked the considerable evidence for genetic influences on intellectual development (Jensen, 1969; Zigler, 1970).

PERSONAL HISTORY OF THE MILWAUKEE PROJECT

Very few second-hand accounts of the Milwaukee Project longitudinal investigation provide any discussion of the cross-sectional research findings on which it was based. The Milwaukee Project is generally misrepresented as an attempt to raise IQ, and as a logical refutation of Jensen's (1969) position that intelligence is highly heritable and therefore not easily changed. It is also misrepresented as an attempt to improve the school performance of children living in poverty.

The Milwaukee Project investigation arose from Heber's earlier work on preparation of the fifth edition of the Manual on Terminology and Classification in Mental Retardation (Heber, 1959). This included an attempt to develop a more refined classification of intellectual retardation with unknown etiology. It was hypothesized that mild and borderline retardation not associated with disease or obvious expressions of pathology was not homogenous, but may have a number of different but unspecified etiologies, which included psychogenic or psychological factors that could interfere with normal development.

Heber was particularly interested in "cultural–familial" mental retardation. This was quite carefully crafted to be a descriptive classification, and there was no "intent to specify either the independent action or the relationship between cultural and genetic factors" in etiology (Heber, 1959, p. 40). This term simply reflected the disproportionate incidence of mild and borderline intellectual retardation within low socioeconomic groups associated with similar intellectual deficits in other family members. He recognized, by virtue of the definition he helped to develop and the design of the Milwaukee Project program of investigation, that research was required to specify the etiology of the mild form of intellectual retardation for which there was no clinically identifiable expressions of central nervous system pathology.

The environmental bias in attempts to explain nonorganic intellectual retardation has not dissipated over the years. A revision of the official definition of mental retardation with unknown etiology (Grossman, 1973) no longer included the term "cultural–familial mental retardation." This category was labeled "retardation due to environmental influences," and included only references to "psychosocial disadvantage" and "sensory deprivation." This nomenclature, which was retained in a later revision (Grossman, 1983), "gives the clear impression that mental retardation without known organic etiology is the exclusive product of a poor environment" (Zigler & Hodapp, 1986, p. 82).

The Milwaukee Project began with an epidemiologic study in the most economically disadvantaged area of Milwaukee. Although all retardation without known organic causes was attributed to environmental factors, the Milwaukee project study was initiated as an hypothesis that such factors contribute to the disproportionate incidence of mild and borderline retardation with unknown etiology within this area. In the early 1960s, a research team from the University of Wisconsin-Madison established a high-risk population laboratory in nine contiguous census tracts of the most economically disadvantaged areas of Milwaukee. These tracts, although containing about 2.5% of the city's population, accounted for 33% of all children labeled "educable mentally retarded" by the public schools.

It was evident from our early work in the central city of Milwaukee that most children living in poverty were not mentally retarded. Rather than investigating factors that obviously differentiated these census tracts from the general population, an effort was made to identify demographic markers characterizing families of children exhibiting more adequate intellectual ability and families of children exhibiting less adequate ability within this subgroup. Flynn (1994, this volume) mischaracterizes the Milwaukee study as an attempt to identify distinctive features that differentiate upper-class and middle-class homes from lower-class homes.

Screening in the nine census tracts identified 88 families in which there were infants less than 10 months old and at least one child of school age. Mothers in these families were administered the Wechsler Adult Intelligence Scale (WAIS), and language samples were collected in an experimental setting. All children in these families were tested using age-appropriate measures of intelligence. A preliminary analysis was conducted on a subsample to identify "low-risk" and "high-risk" families (Aserlind, 1963). Thirteen families whose oldest child scored 85 or higher were designated "low-risk," and 13 families whose oldest child scored 84 or less were designated "high-risk." The intellectual and language data samples of the mothers in these families were then compared. The mean IQ of the "low-risk" mothers was significantly higher than the mean for "high-risk" mothers, and their language samples were more sophisticated in terms of mean length of verbal response, total words used, and the use of connectives. In addition, the mean IQ of the oldest children in the "high-risk" families approximated the mean IQ of their mothers, but the mean IQ of the oldest children in the "low-risk" families was significantly higher than the mean IQ of their mothers.

The findings of this analysis suggested that maternal IQ may be an effective risk index for low IQ in children in these families. To further test this relation, scores for the entire sample of 88 families were analyzed. The mean IQ of all children in the sample (excluding newborns) was 86.3, and the prevalence of scores of 75 or lower was 22%. These data showed several important characteristics of this population group, including the fact that the prevalence rate was considerably higher than the generally accepted rate of 3% estimated for the larger population. This indicated that there were, in fact, a disproportionate number of intellectually retarded children in the families of this population subgroup. However, we also found that these retarded children were not evenly distributed among the 88 families. Families were divided into two groups, those with mothers scoring above 80 (the mean maternal IQ of 80.5 rounded) and those with mothers scoring below 80. The 45% of the families with mothers scoring below the mean contained 78.2% of the children who scored below 80. This disproportionate relation between maternal and children's IQs was even greater at lower levels of maternal IQ.

It is most important to emphasize that the major and original interest in this investigation was not simply the low IQs of children, but the decline in IQ across age. We were trying to determine if there was evidence which would indicate that the rate of intellectual development may have been depressed. To investigate this possibility, means were calculated for all children in specific age categories and plotted separately for "high-risk" and "low-risk" families. The mean IQ was relatively comparable across age categories for children whose mothers scored above 80, but there was a progressive decline in mean IQ across age categories for children who mothers scored below 80.

In addition to evidence for declines in the rate of intellectual development across age, the survey results were also expected to provide an effective index for identifying normally functioning infants who were most at risk for declines. The first analysis indicated that low IQs were not evenly distributed or generally characteristic for children in this economically disadvantaged population. In addition, the cross-sectional analysis of scores across age categories suggested that the decline in IQ was limited to children in families with mothers scoring relatively low themselves. The risk for such declines was not adequately defined by the general conditions of poverty, but could be more specifically targeted within families in which parents had relatively low IQs. Risk was therefore familial rather than social or cultural as had been assumed previously.

We recognized that the familial risk identified in this survey could represent the influence of genetic factors, environmental deprivation, or the combined influence of these factors. Clinical observations made in the homes when intellectual tests were given suggested that the mothers who scored lowest created a social environment for their children that was distinctly different than the social environment created by higher functioning mothers living in the same general disadvantaged area and conditions. Although no effort was made to formally test this difference, it was hypothesized that the experience of being raised by a primary caregiver who is mentally retarded was not adequate to support development within the reaction range defined by the genetic determinants of intelligence, particularly during infancy when children are almost totally dependent on others to care for them and to structure their environment for them.

The general findings of this survey research have consistently been reported in some form in every account of the Milwaukee investigation. It is difficult to understand how anyone could read these accounts and conclude that the longitudinal investigation, which was based on these findings, was an attempt to study the influence of poverty on intellectual development or an attempt to raise IQ.

The longitudinal experimental investigation was designed to test the hypothesis that there is causal relation between the experience of being raised by a retarded mother and declines in IQ across age. This prospective research effort focused on the prevention of delays in intellectual development for normally functioning infants through manipulation of their early experience (Heber & Dever, 1970; Heber, Dever, & Conry, 1968).

Intervention began when the children were less than 6 months of age, and all children were functioning within the normal range. At 10 months of age there was less than a one-half month difference in the mean mental age of the treated and untreated children as measured by the Gesell Schedules of Development. Treated children maintained a relatively stable mean DQ through 18 months of age, but then demonstrated a 5 to 6 point gain in mean DQ between 18 and 22 months of age, and an additional 3-point gain in mean IQ as measured by the Stanford Binet at 36 months of age. This 8- or 9-point rise in mean DQ and IQ does not represent a real increase in intelligence, but has been attributed to artificial inflation of scores, which is inevitable in any intervention program (Garber & Hodge, 1989; Garber et al., 1991).

The most startling finding of this investigation was not the 8- or 9-point rise in the mean DQ and IQ of treated children, however, but the rapid and significant decline in the mean DQ and IQ for control children who stayed at home with their retarded mothers. After correcting for the artificial inflation of scores of the treated children, Garber and Hodge (1989) estimated that the true intervention effect was the prevention of decline of about 22 IQ points for experimental children.

The decline in IQ scores of children in this population occurred prior to 3 years of age and differed substantially across individuals. There was a significant relation between the magnitude of decline and the original DQ of each child at 10 months of age. Children with the highest original scores demonstrated the greatest decline across age. The experience of being raised by a retarded mother must therefore be defined in terms of transactions between the genetic potential of each child and the intellectual stimulation provided by their mother.

THE DEVELOPMENT OF A MYTH

We reiterate our concern that it is difficult to understand how anyone could read original accounts of this research and conclude that the longitudinal investigation was an attempt to raise IQs or to refute Jensen's (1969) position that intelligence is highly heritable, or that it was a study of the influence of poverty on intellectual development. What, then, was the origin of this myth that has surrounded the Milwaukee Project almost from its beginning?

The process by which second-hand and third-hand sources come to err in their descriptions of often-cited studies may be no more than a simple drift away from the original due to retelling or submission to the convenience of explanation offered by a particular theoretical scheme. There is, however, also the more dynamic influence on the initial retelling of research due to the authors own theoretical bias, and the need to establish a point (Cornwell & Hobbs, 1976; Harris, 1979). Myth making need not be viewed as a conscious attempt to deceive, but rather as an attempt to build a sense of cohesiveness and continuity in a particular area of study. This requires providing unambiguous examples of theoretical positions, and often involves minimizing differences, overgeneralization of similarities across studies, and engaging in semantic reversals in descriptions. Such bias occurs in both efforts to support a position and efforts to refute a position and arise largely as a product of pedagogy.

Blumenthal (1975) noted that research is carried out and interpreted in a stream of history, both individual and universal. The theories that dominate an area of research during specific periods will create general biases in how research is described and interpreted, regardless of the original intent of the research. During the early 1960s, when the Milwaukee Project was being planned and developed, there was increased concern about the general academic achievement of all children, but particularly about the relatively poor achievement of children living in the most economically disadvantaged areas of our society. The failure of these children to achieve was generally attributed to the fact that when they entered school their average intellectual level was significantly lower than the average for the general population of children entering school. As was noted in the discussion of the Milwaukee Project, the decline in IQ across age, which was considered typical for these children, was attributed almost uncritically to depression in the rate of intellectual development by the general public.

Noting the failure of early intervention programs to produce meaningful and lasting increases in IQ for children living in poverty during the late 1960s, Jensen (1969) expressed concern with the almost total emphasis on the concept of "social deprivation" as an explanation for all individual differences in intelligence. Contrary to what some believe, he did not deny that experience influences intelligence. However, he did argue that, with exception of the rare few children born with neurological deficits, the "average child" concept viewed all low scoring children as basically much alike, and he attributed differences in intellectual development for these children to what he considered rather superficial differences. Attributing all but small differences to differences in experience was naive. He believed the considerable evidence for genetic influence on intelligence was being ignored.

Edward Zigler (1970), a recognized leader in the field of mental retardation and the planner of Head Start, also argued that the evidence of environmental influences on intelligence was being overgeneralized. The belief that borderline and mild intellectual retardation not associated with organic pathology was due to environmental influences, the theoretical bias for most social programs and early intervention research, was simply unrealistic.

Unfortunately, these warnings were not heeded. The belief that poverty caused all nonorganic retardation became more entrenched when Jensen (1969) rekindled the nature–nurture debate by suggesting that the failure of early intervention programs to raise and maintain IQs was due to the fact that intelligence was highly heritable and not easily changed. This resulted in an increasing need to provide narrowly defined unambiguous examples in support of the theoretical position that the environment influences intelligence, and led to the minimization of differences and overgeneralization of similarities across studies.

The early results of the Milwaukee Project provided some of the best evidence that early experience influences the rate of intellectual development. Unfortunately, those with an environmental bias failed to recognize that this investigation was actually a challenge of their extreme views. The substantial differences between this prospective intervention and the majority of early intervention efforts that attempted to raise the IQs of all children living in poverty were not recognized. These individuals concentrated on the fact that the children in the Milwaukee Project lived in poverty and ignored the major finding that the risk for declines in IQ was limited to children being raised by retarded mothers.

Critics, beginning with Page (1972), have also misinterpreted this research as an attempt to raise IQ. As we noted previously, early accounts of this investigation contained inadequate detail, but Page, we believe, had a more immediate concern that the laudatory commentary by the media would, in effect, be seen as a final rebuttal to hereditarians without sufficient proof. He did note that Heber had personally assured him before any published criticism appeared that the Milwaukee Project was in no sense a test of Jensen's position that intelligence had high heritability and was not easily changed. Page asserted, however, that it was reasonable for environmentalists to assume that it was actually an attempt to discredit Jensen simply because it addressed the question of an environmental influence on intelligence. Jensen (Personal Communication, 1991) has acknowledged that the Milwaukee Project research is in no way inconsistent with his position.

A recent example of the continuing denial for the possibility of a genetic basis for nonorganic retardation can be found in Jackson (1993). Paradoxically, the author cites the results of the Milwaukee Project as evidence that it may be possible to raise the IQs of African American children living in poverty. This article references the final report of the research (Garber, 1988) and two of our responses to critiques of this report (Garber & Hodge, 1989; Garber, et al., 1991). Unfortunately, her personal bias and need to support a theoretical position have interfered with an accurate interpretation and representation of our conclusion about the major findings of this research.

THE MYTH OF REPLICATION

The increased emphasis on improving the school performance of children living in poverty led to new research in the early 1970s. The Abecedarian Project (Ramey & Campbell, 1987; Ramey & Haskins, 1981) was designed to prevent school failure for children living in a rural setting. Like the Milwaukee Project, this was a prospective investigation that attempted to prevent the decline across age by providing stimulation outside the home. This, however, was essentially the only similarity between these two research efforts. Unfortunately, one of the most common distortions in current reviews of early intervention literature is the characterization of the Abecedarian Project as a replication of the Milwaukee Project.

Garber and Hodge (1991) noted that these two research efforts were actually based on very different conceptions of the influence of the environment on intelligence. The Milwaukee Project defined the risk for declines in IQ in terms of the experience of being raised by a retarded mother, but the Abecedarian Project defined risk in terms of social disadvantage. "The Abecedarian Project, which was begun after the Milwaukee Project intervention was over, chose to ignore the finding that the risk for declines in IQ across age are familial rather than social" (p. 321).

The first published discussion of the theoretical basis of the Abecedarian Project (Ramey et al., 1976) described the Milwaukee Project as "without question the most relevant project to the issue of preventing mental retardation conducted thus far" (p. 632). These authors went on to provide a detailed description of the early survey reported in Heber, Dever and Conry (1968), and the major finding that:

> It is not just the "poor" or "lower classes" who contribute the "cultural–familial retardate," it is certain families belonging to a certain group within this population who make the largest contributions. It is a relatively small percentage of families within the deprived economic group which contributes very heavily to the high prevalence of "cultural–familial" retardation. (Ramey et al., 1976, p. 632)

Their description of the Milwaukee survey results supports the assertion that they knew the risk for declines in IQ across age was better defined in terms of low maternal IQ than socioeconomic status. However, when it came time to select the sample for their investigation, they chose to use an experimental version of a 13-item High-Risk Index comprised primarily of measures reflecting social disadvantage.

Ramey and Haskins (1981) reported that the mean IQ of mothers in the Abecedarian Project was 84. This is about 17 points higher than the mean IQ of mothers in the Milwaukee Project. The importance this difference makes for the selection of children at risk for declines in IQ is best illustrated by the fact all children in the Milwaukee Study would have qualified for inclusion in the Abecedarian Project, but less than 15% of the children in the Abecedarian study would have qualified for inclusion in the Milwaukee Project study.

A NEW MYTH

Over the years, the reports on the Abecedarian Project, and the complimentary Project CARE, have consistently emphasized that these were studies of the influence of poverty on intellectual development and achievement. However, several recent unpublished reports (Ramey & Landesman, 1989; Ramey & Landesman-Ramey, 1992; Ramey & Ramey, 1990, 1991) and an article by Martin, Ramey, and Ramey (1990) began reporting what are referred to as "new findings."

A post hoc review and comparison of the subsample of children with mothers who had IQs of 70 or lower with children with mothers who had IQs of 95 or higher indicated that the educational intervention was particularly positive for children with low-IQ mothers. By 54 months of age there was a 22-point difference between the mean IQ of experimental program children of retarded mothers and the control group children of retarded mothers.

Spitz (1992) noted that this subgroup of children with retarded mothers contributed disproportionately to the difference identified for the entire Abecedarian population. He suggested that these results seem to support Garber et al.'s (1991) contention that the risk for declines in IQ resulted from the experience of being raised by a primary caregiver who is retarded rather than from being raised in a poor socioeconomic environment.

Unfortunately, the finding that the single strongest predictor of declines in IQ is the mother's level of tested intelligence is being reported as a "new finding." If these authors mean this is a new finding within their own data, they are, of course, correct. But Ramey and Landesman (1989) state that, "The findings from our study are the first, to our knowledge, to demonstrate positive disruptions, on an individual family basis (i.e., mother–child pairs), to expected intergenerational patterns of low intelligence" (p. 11). In any event, the findings of the Milwaukee Project, which demonstrated this 20 years earlier, and of which these authors are well aware, are never mentioned in any of these reports.

CONCLUSION

Advancement of knowledge in any area of the psychology of human behavior requires consideration of evidence in support of competing theories for which it is presumed that interpretation has been based on honest representations of the purpose, procedure, and outcome of research. We believe that the factors discussed here are, among others, contributors to delays in developing a more useful understanding of both intelligence and intellectual development. There also remains an especially elusive concern; namely, how to minimize the tradition of confounding definitions of intelligence either for purposes of "interpretational convenience" or confusion through lack of understanding, particularly where one theoretical system is not clearly related methodologically to a research problem and leads to misinterpretation. Some help is offered by Sternberg (1988), who has proposed a schema for conceptions of intelligence that students of the problem can use to clarify both their research approach and their understanding of other's research.

Discussions of the intellectual development of children identified to be at risk for intellectual delay and selected to participate in early childhood programs investigating the determinants of intellectual development in such children have mainly featured psychometric indices as behavioral equivalents by which to ascertain the success or failure of such programs. This approach reinforces an esoterically narrow view of intellectual development as predominantly a research problem of psychometrics rather than the parallel research problem concerned with the intellectual development of children at risk as a phenomenon of the interaction between an individual and his or her environment. The emphasis on traditional psychometric data as the predominant basis for the interpretation of results from studies of development masks the importance of individual differences and minimizes the potential impact of the unique psychosocial context of at risk children, who are most often only simply described as poor, disadvantaged, or low SES.

Though it has been hoped that these parallel lines of psychometric research of intelligence and early intellectual development of individuals in interaction with their environment would one day usefully cross, the blurring of these research streams continues to confuse both. The result has been to seriously delay effective reform of early education strategies for children at risk, for at least the last quarter century. Although there is no requirement that the research be useful, there is, however, in the relationship of the study of intelligence to the issue of children's intellectual development, matters of considerable social urgency that would likely benefit.

EPILOGUE: IN THE MEANTIME

It was a cold evening in December, 1990, and at 4:00 p.m. in Milwaukee, the sky, already a winter gray, was beginning to darken. We had spent most of the day working on a manuscript (Garber, et al., 1991) describing the history of the Milwaukee Project and making our argument for what we thought had been our contribution to the understanding of the intellectual development of children at risk. In a moment of nostalgia we decided to visit the first site of the Milwaukee Project's infant stimulation program; a gray, three-story duplex that had been refurbished just for the project in the neighborhood from which the children and their families were drawn for the study. We arrived there in less than 10 minutes from the center of the city, finding the site with surprising ease after so many years. What we saw caused the five of us to become silent—absolutely silent. Later we realized that we had all seen and thought the same thing. Since 1967, nearly 25 years had passed and the building hadn't changed, the neighborhood hadn't changed, and, most unfortunately, neither had the lives of the children from this community. In fact, conditions were worse than before. More children in poverty, more unwed and younger mothers with more children under school age, and more social and economic despair.

The shock of that moment came from the fact that as we all reflected on our experience we all thought that somehow something had happened over the last 25 years that we could see here. After all, hadn't we, during those 25 years, written many papers, presented at many conferences, and made our many academic arguments, and, if you will forgive our cynicism here, in spite of all that effort, shouldn't we have been able to feel, on that slowly darkening street in Milwaukee, that somehow some good was happening for children? For us, as we talked, it was clear that there was hope that good things will yet happen, but perhaps only if more of us see not only the children's performance data but also see the children!

REFERENCES

Aserlind, L. (1963). An investigation of maternal factors related to the acquisition of verbal skills of infants in a culturally disadvantaged population. *Dissertation Abstracts International*, 3197.

Blumenthal, A. L. (1975). A reappraisal of Wilhelm Wundt. *American Psychologist, 30*, 1081–1088.

Brand, D. (1987, August 31). The new whiz kids: Why Asian-Americans are doing so well, and what it costs them. *Time*.

Bringmann, W. G., Balance, W., & Evans, R. B. (1975). Wilhelm Wundt 1832–1920: A biographical sketch. *Journal of the History of the Behavioral Sciences, 11*, 287–297.

Butterfield, E., & Berkson, G. (1991). Special book review section: The Milwaukee Project clarified (Reviews by D. Guthrie, D. MacMillan, K. Widaman, H. Spitz, & R. Weinberg). *American Journal on Mental Retardation*, *95*, 477–492.

Cornwell, D., & Hobbs, S. (1976). The strange saga of little Albert. *New Society*, *18*, 602–604.

Ellenberger, H. F. (1972). The story of "Anna O." A critical review with new data. *Journal of the History of the Behavioral Sciences*, *8*, 267–279.

Flynn, J. R. (1984). The mean IQ of Americans: Massive gains 1932 to 1978. *Psychological Bulletin*, *95*, 29–51.

Garber, H. L. (1988). *The Milwaukee Project: Preventing mental retardation in children at risk*. Washington, DC: American Association on Mental Retardation.

Garber, H. L., & Hodge, J. D. (1989). Risk for deceleration in the rate of mental development. *Developmental Review, 9*, 259–300.

Garber, H. L., Hodge, J. D., Rynders, J., Dever, R., & Velu, R. (1991). The Milwaukee Project: Setting the record straight. *American Journal on Mental Retardation, 95*(5), 493–525.

Goodwin, C. J. (1991). Misportraying Pavlov's apparatus. *American Psychologist, 104*, 135–141.

Grossman, H. J. (1973). *Manual on Terminology and Classification in Mental Retardation*. American Association on Mental Deficiency.

Grossman, H. J. (1983). *Manual on Terminology and Classification in Mental Retardation*. American Association on Mental Deficiency.

Harris, B. (1979). Whatever happened to little Albert? *American Psychologist, 34*(2), 151–160.

Heber, R. F. (1959). A manual on terminology and classification in mental retardation. *American Journal of Mental Deficiency, 65*, (Monograph Suppl.).

Heber, R. F., & Dever, R. B. (1970). Research on education and habilitation of the mentally retarded. In H. C. Haywood (Ed.), *Social–cultural aspects of mental retardation* (pp. 395–427). New York: Appleton-Century-Crofts.

Heber, R. F., Dever, R. B., & Conry, J. (1968). The influence of environmental and genetic variables on intellectual development. In H. J. Prehm, L. A. Hamerlynck, & J. E. Crossen (Eds.), *Behavioral research in mental retardation* (pp. 1–22). Eugene: University of Oregon Press.

Jackson, J. F. (1993). Human behavioral genetics, Scarr's theory, and her views on interventions: A critical review and commentary on their implications for African American children. *Child Development, 64*, 1318–1332.

Jensen, A. R. (1969). How much can we boost IQ and scholastic achievement? *Harvard Educational Review, 39*, 1–123.

Jensen, A. R. (1989). Raising IQ without increasing *g*? A review of the Milwaukee Project: Preventing mental retardation in children at risk. *Developmental Review, 9*, 234–258.

Locurto, C. (1988). On the malleability of IQ. *The Psychologist: Bulletin of the British Psychological Society, 11*, 431–435.

Locurto, C. (1991). Beyond IQ in preschool programs? *Intelligence, 15*, 295–312.

Locurto, C. (1991). Hands on the elephant: IQ, preschool programs, and the rhetoric of inoculation—A reply to commentaries. *Intelligence, 15*, 335–349.

Martin, S. L., Ramey, C. T., & Ramey, S. (1990). The prevention of intellectual impairment in children of impoverished families: Findings of a randomized trial of educational day care. *American Journal of Public Health, 80*(7), 844–847.

Page, E. B. (1972). Miracle in Milwaukee: Raising the IQ. *Educational Researcher, 1*, 8–15.

Page, E. B. (1986). The disturbing case of the Milwaukee Project. In H. H. Spitz (Ed.), *The raising of intelligence: A selective history of attempts to raise retarded intelligence* (pp. 115–140). Hillsdale, NJ: Erlbaum.

Page, E. B., & Grandon, G. M. (1981). Massive intervention and child intelligence: The

Milwaukee Project in critical perspective. *Journal of Special Education, 15*, 239–256.

Paul, D. B., & Blumenthal, A. L. (1989). On the trail of little Albert. *The Psychological Record, 39*, 547–553.

Prytula, R. E., Oster, G. D., & Davis, S. F. (1977). The "rat rabbit" problem: What did John B. Watson really do? *Teaching of Psychology, 4*, 44–46.

Ramey, C. T., & Campbell, F. A. (1987). The Carolina Abecedarian Project: An educational experiment concerning human malleability. In J. J. Gallagher & C. T. Ramey (Eds.), *The malleability of children* (pp. 127–139). Baltimore, MD: Brooks Publishing.

Ramey, C. T., Collier, A. M., Sparling, J. J., Loda, F. A., Campbell, M. D., Ingram, D. L., & Finkelstein, N. W. (1976). The Carolina Abecedarian Project: A longitudinal and multidisciplinary approach to the prevention of developmental retardation. In T. D. Tjossem (Ed.), *Intervention strategies for high-risk infants and young children* (pp. 629–665). Baltimore, MD: University Park Press.

Ramey, C. T., & Haskins, R. (1981). The causes and treatment of school failure: Insights from the Carolina Abecedarian Project. In M. Begab, H. Garber, & H. C. Haywood (Eds.), *Causes and prevention of retarded development in psychosocially disadvantaged children*, 2 (pp. 89–125). Baltimore, MD: University Park Press.

Ramey, C. T., & Landesman, S. (1989). *Prevention of intergenerational intellectual retardation.* Paper presented to the International Association for the Scientific Study of Mental Retardation, Dublin, Ireland.

Ramey, C. T., & Landesman-Ramey, S. (1992). Effective early intervention. *Mental Retardation, 30*(6), 337–345.

Ramey, C. T., & Ramey, S. L. (1990). *Prevention of intergenerational mental retardation.* Paper presented at the annual meeting of the American Association on Mental Retardation, Atlanta.

Ramey, C. T., & Ramey, S. L. (1991). *Mothers with mental retardation: Enhancing their development and that of their children.* Paper presented at the annual meeting of the American Association on Mental Retardation, Washington, DC.

Samelson, F. (1974). History, origin, myth and ideology: "Discovery" of social psychology. *Journal for the Theory of Social Behavior, 4*, 217–231.

Spitz, H. H. (1992). Does the Carolina Abecedarian Early Intervention Project prevent sociocultural mental retardation? *Intelligence, 16*, 225–237.

Sternberg, R. J. (1988). *The triarchic mind: a theory of human intelligence.* New York: Penguin.

Todd, J. T., & Morris, E. K. (1992). Case histories in the great power of steady misrepresentation. *American Psychologist, 47*(11), 1441–1453.

Vernon, P. E. (1982). *The abilities and achievements of Orientals in North America.* New York: Academic Press.

Zigler, E. (1970). The nature–nurture issue reconsidered. In H. C. Haywood (Ed.), *Social–cultural aspects of mental retardation: Proceedings of the Peabody–NIMH conference.* New York: Appleton-Century-Croft.

Zigler, E., & Hodapp, R. M. (1986). *Understanding mental retardation.* New York: Cambridge University Press.

Chapter 12

Secular Trends in IQ: Additional Hypotheses

Arthur R. Jensen

School of Education, University of California, Berkeley

The causes of the secular gains in IQ reported by James Flynn (Chapter Two) are still a mystery, or, at best, unproved hypotheses. At this stage it would seem wise to broaden the scope of hypotheses regarding plausibly causal variables and try to examine their consistency with the available evidence.

The situation with IQ measurements seems to be analogous to measuring the people's heights without using a ruler (which we assume gives the true measurement), but by measuring the lengths of the shadows they cast in a particular location at a particular time of day. The shadow measurements would of course be correlated with true height as much as reliability would permit, and the shadow measurements thus obtained on people at the same time and place would validly predict other variables that are correlated with height, such as leg length, shoe size, and weight. If the identical twins of all the people in this group had their shadows measured at a different time of day (or at a different latitude), however, not only would the mean heights of the two groups differ, but if the shadow measurements of the two groups were mixed together they would have a lower correlation with true height and would also lose in predictive validity. If we had some way to know the true heights, we could devise a set of conversion equations for the shadow measures obtained at different times and places, making them all equally correlated with true height and equally valid predictors. I have suggested elsewhere (Jensen, 1991) how this might be done to deal with the problem of intergenerational population trends in IQ, using minimal information-content measures of information-processing capacity or physiological measures of brain activity. Anchoring conventional psychometric scores to such process measures would help in determining how much of the secular trend is attributable to test-taking artifacts (what Flynn refers to as the "Brand hypothesis") and to actual changes in the biological substrate of intelligence, such as the effects of improved nutrition (the "Lynn hypothesis").

This shadow analogy is consistent with the fact that, unlike the population raw-score means of IQ tests, the variance, the reliability of tests, the pattern of intercorrelations among various tests, their factor structure, their predictive validity, and correlations with many nonpsychometric variables, the correlations of twins and other kinships, their heritability coefficients correlations, and their construct validity remain practically constant across generations.

Which factors affecting the whole population could bring about an appreciable trend in the test scores' central tendency without any corresponding changes in all these other statistics? It would have to be something that has acted upon virtually the whole populations of industrialized countries in modern times.[1] A number of things have happened in the last 50 or 60 years that have affected these populations, some things affecting only test scores per se ("shadow" measurements) and some affecting the biological substrate with which all highly g-loaded test scores are correlated. We have probably underestimated the part of the population gains in IQ that are attributable to the biological correlates of psychometric *g*.

The "shadow" aspect of IQ gains is probably the result of such historically recent effects as universal public education, the increasing use of psychometric tests (especially of the objective and multiple-choice variety), general test sophistication, changes in attitudes, guessing tendencies, and the like, and increasing popular emphasis on early childhood cognitive development via

---

[1] It is a mistake to extrapolate the trend in mental test scores, either backwards or forwards in time, beyond the limits of the period actually covered by the existing normative data. Although astronomers are able to look back in time to study the history of the universe, based on the postulate that the speed of light is constant throughout time and space, such linear extrapolations of population trends often lead to absurdity. For instance, if we assume that Aristotle, Shakespeare, and Newton all had IQs of 200 in their day, and if, from the dates of their greatest achievements, we extrapolate ahead, using Flynn's claimed in 15 IQ-points average gain per generation in the general population, they, if revived and tested today on a culture-fair IQ test (say, Raven's matrices) using present-day norms, would have respective IQs of –973, 0, and 46!

radio, television, nursery schools, children's books, and games. These are the kinds of psychological and educational influences that raised children's IQ scores some 20 points without increasing *g* in the famous Milwaukee Project (Jensen, 1989).

The true aspect of IQ gains may be attributable to the summation of a great many possible causes, each of which alone has but a very small but real effect on mental development. The same factors are reflected also in the secular increase during this century in growth rates and in adult physical stature, earlier age of menarche, decreased rates of fetal loss and infant mortality, and increased longevity. The population trends in these variables are fairly paralleled by the trend in test scores. The gain in height, for example, has been about the same as for IQ, when both variables are measured in standard deviation units.

Here are some of the hypothesized causal factors deemed largely responsible for these trends. All of them have been found to be significantly correlated with IQ. Improved nutrition has affected virtually the entire population of First World countries through mass production and general availability of processed foods, which are often fortified with vitamins and minerals. Widespread improvements in health care have included marked decreases, by nearly universal inoculation, in the incidence of many previously common childhood diseases, each of which probably takes a small toll on mental growth. Also, a host of various prenatal and perinatal factors are known to be correlated with IQ. Advances in obstetrical methods have greatly lessened the incidence of the prenatal and perinatal risks known to result in "reproductive casualties" that often affect subsequent mental development. With the decreasing family size in industrialized countries, there is naturally a decrease in parity (i.e., number of previous births by a given mother), a variable that is negatively correlated with IQ. Some forms of blood-type incompatibility between mother and fetus, as in kernicterus, where the mother is Rh negative and the fetus is Rh positive, are negatively correlated with the child's later IQ. Such immunoreactive effects have also been found in the ABO blood groups, and there are probably other maternal antigens among the many microenvironmental factors that affect fetal development, with later effects on IQ. The increasing use of Rh-immune globulin in second and subsequent pregnancies during the last two generations has greatly diminished the effects of these immunoreactive factors. (References to this literature and more detailed discussion of its implications for IQ may be found in Jensen, in press.) Also, increases in migration and people's increasing social mobility with the breakdown of class barriers have resulted in more outbreeding, which has a heterotic, or hybrid vigor, effect on IQ, as it has on stature and other physical traits. The increased use of electric lighting and TV viewing has also probably speeded up the rate of maturation in humans, via stimulation of the pineal gland, as well as in animals in which this effect has been demonstrated experimentally.

But, as Flynn notes, it is still quite unknown just how much of the secular increase in scores on *g*-loaded tests is due to the "shadow" aspect of mental measurement and how much is due to real changes in the biological substrate of mental development. It is important for the theory of intelligence that the answer be found. I suspect, though, that the answer will have to await the development of direct measures of the brain mechanisms involved in *g* that can be obtained on this and subsequent generations (see Jensen, 1991). If the upward trend in the population's average level of performance on *g*-loaded psychometric tests in First World countries has resulted from factors that now are near the asymptote of their positive effect on mental development, we will not be able to discover the answer to Flynn's question within the populations of the industrialized nations. The answers will have to be sought in future intergenerational studies conducted in Third World countries that are just beginning to see the effects of modern industrialization and all its consequences.

REFERENCES

Jensen, A. R. (1989). Raising IQ without increasing *g*. *Developmental review, 9*, 234–258.

Jensen, A. R. (1991). Speed of cognitive processes: A chronometric anchor for psychometric tests of *g*. *Psychological Test Bulletin, 4*, 59–70.

Jensen, A. R. (in press). The puzzle of non-genetic variance. In R. J. Sternberg & E. L. Grigorenko (Eds.), *Intelligence, heredity, and environment*. New York: Cambridge University Press.

Chapter 13

Environment and Intelligence: A Comment

John C. Loehlin

The University of Texas at Austin

Reading these seven chapters on intelligence and the environment was a rather soul-stressing experience for a person of simple intellectual tastes. My goodness, it's all terribly complicated. The environment is a huge collection of interacting systems and supersystems and subsystems specialized into more-or-less distinct domains of culture and expertise, with every system interacting both ways with every other system, and all of this changing continuously over the course of development and training and practice—and that's only the beginning. We also have to take into account the multilevel biological subsystems within every individual and how they interact with the environmental systems, with each other, with their genetic specifications, and with their nutritional inflows.

I have no doubt whatever that all of this complexity is in some sense true, and that we need always to keep it at least somewhere in the back of our minds. And yet amid this welter of interactions, it is important to remember that there are main effects—big ones—that can orient us as we obsess over environment–intelligence relationships in some specialized corner of this vast enterprise. What are some of these empirical beacons in the night? Here are a few:

1. Identical twins reared apart correlate on the average about .74 in their performance on IQ tests (Bouchard, Lykken, McGue, Segal, & Tellegen, 1990).
2. Genetically unrelated children reared together in the same family from near birth until late adolescence correlate essentially zero in their performance on IQ tests (Loehlin, Horn, & Willerman, 1989; Scarr & Weinberg, 1978).
3. The correlation between the Vocabulary and Block Design subscales of the WAIS-R is about .52 for the U.S. standardization population (Wechsler, 1981).
4. Average IQs have been rising worldwide in developed countries at the rate of some 10 to 20 points per generation for nonverbal, and 5 to 10 points for verbal tests (Flynn, Chapter 2).

What do these astonishing facts tell us about intelligence and environment? How might this apply to some of the topics addressed in the seven chapters on intelligence–environment relationships?

Fact #1 says that individual differences in performance on IQ tests (in the relevant populations) are mostly genetic in origin. (The alternative hypothesis fancied by some, that a substantial chunk of this correlation may stem from the fact that separated identical twins are sometimes placed in similar families, is rendered unappealing by Fact #2). Now, Fact #1 doesn't say that environments aren't important—obviously, without them intelligence wouldn't develop at all. And it doesn't say that gene–environment correlations of the active and reactive sorts can't be involved (Plomin, DeFries, & Loehlin, 1977). Some of the resemblances between separated twins may result from the fact that other people react to them in similar ways because their appearance and behavior are alike, or that the twins themselves may seek out similar learning experiences. Some of the ways in which genes express themselves in any animal species involve behavior. Beaver genes don't automatically get transcribed into a beaver dam—the beavers have to go out and fell trees and move the logs to the site of the dam and arrange them properly. Any of these steps may be modulated according to the details of the terrain. Yet the result is a structure characteristic of the species (Dawkins, 1982). Fact #1 probably does say, though, that gene–environment *interaction* (in the statistical sense of the same genes having a different effect in different environments) can't be hugely important quantitatively in accounting for IQ variation. After all, GE interaction has to divide up the leftover 26% with errors of measurement, with developmental accidents, and with any systematic effects of environmental variables that differ between the twins—all factors that presumably contribute appreciably to IQ variance.

Fact #2 is based on a somewhat restricted range of environments, but within that range its message is clear: The environmental variables that the children in a family share don't make them alike in IQ in the long run. (In the shorter run, there *is* an effect, as suggested by various correlations of .11 to .49 among adoptive siblings measured in childhood—Loehlin et al. 1989; Scarr & Weinberg, 1977). Think of some of the environmental variables on which the children in a family are typically more alike than are randomly paired individuals from the population. They will include such popular favorites in the environment–IQ game as the number of newspapers, books, and magazines in the home, the affluence of the neighborhood in which the children grow up, the quality of schools they attend, the parents' aspirations for their childrens' education, the parents' personalities and attitudes toward childrearing, the warmth and cohesion of the family—the list goes on and on. All of these things *can* differ from child to child within a family, but they should tend to differ more for children reared in different families, and thus, if they are important influences on IQ, should produce a correlation among genetically unrelated individuals reared together. They do not.

Fact #3 speaks to the generality of intellectual development. The two tasks themselves have, on the surface, almost nothing in common, stating the meanings of words and copying designs with blocks. One is verbal, one is spatial; one is speeded, the other is not; one is fluid, one crystallized, in Cattell's (1971) classification; one achievement is a central object of conventional schooling, the other is not. And yet the two tasks share over half their variance. To the extent that both reflect some common underlying variable (call it intelligence or *g* or neural speed or what you will) they would each correlate .70 or so with that variable. That is a pretty high number, as psychologists' correlations go. Any credible theory of intellectual development has to take this sort of fact into account.

Fact #4 is fully discussed by Flynn in his chapter. It is a large environmental effect, and it is particularly important in reminding us, in conjunction with Facts #1 and #2, that the effect of something on means and on individual differences may not be the same. (*May* not be; whether, in particular cases, it *is* or not must still be demonstrated empirically, a fact sometimes forgotten). Also, as Flynn emphasizes, it makes a lot of difference in interpreting Fact #4 how much of the effect occurs on the measure—the IQ test score—and how much on the underlying trait—intelligence.

How do these four facts help us when we look at some of the particular aspects of environment and intelligence discussed in the main chapters of this book?

What, for example, will they do for the case of attachment and self-regulation, discussed by Borkowski and Dukewich in Chapter 1? In fact, not much, because the authors are quantitatively so modest—less than 10% of IQ variance in children accounted for by attachment, and less than 20% by self-regulation. However, let us be bolder than they, and consider the case for attachment being a major player in explaining IQ differences. What would our four facts suggest? Fact #4 is probably not immediately relevant, unless there is evidence concerning secular trends in attachment (I do not know). From the standpoint of Fact #3, things look just fine. Attachment could have a common effect on quite disparate intellectual tasks. Facts #1 and #2 would suggest an interesting and controversial hypothesis: that if the attachment status of a child has a lasting affect on his or her IQ, that status must largely depend on the child's genes. In that event, identical twins reared in separate families will have similar attachment statuses and thus similar IQs, and genetically unrelated siblings will be uncorrelated in their attachment, and hence IQ. This, by the way, allows for the common finding that insecurely attached parents are likely to have insecurely attached children—the genes that interfered with the parents' secure attachment are now doing the same for their offspring. We even have a way to test this: an appropriately designed adoption study. In the absence of a genetic relationship, neither the attachment statuses nor the IQs of parents should match those of their children.

Let's consider another example: the implications of exceptional performance, as set forth by Wagner and Oliver, Chapter 7. They find that people who are very, very good at something typically have gone through many years of intensive practice at it. They suggest that conventional schooling may represent the many-hours-per-day-over-many-years kind of application required to achieve high performance levels on some of the skills measured by IQ tests. How does this view look from the standpoint of our four facts? From the standpoint of Fact #4, it looks fine. A general increase in years of schooling has been going on during the period of rise in IQ-test performance. However, the lesser improvement in the more schooled aspects of IQ tests (verbal) would still require a bit of fancy footwork. Fact #3 gives us some trouble. Exceptional performance is said to be domain-specific, yet here we have remarkable domain generality to be accounted for. Also, it is a bit awkward that one of the two skills in question is so much more explicitly the focus of training than the other. How about Facts #1 and #2? Here, as in the case of attachment, we would need to assume that the child's genes were the main determinant of the amount of training received, so that identical twins in separate families wound up getting similar amounts, and that genetically unrelated children in the same families did not. This last fact would suggest that it is not sheer exposure to training that matters, but how it "takes"—possibly implicating gene-based personality and motivational factors as well as gene-based capacities for the development of intellectual skills.

Further examples could be elaborated, but I hope that these two cases will begin to illustrate the power of a few key facts to channel theory. Naturally, a hypothesis that passes the test of conformity with these four facts may still be wrong. That something *could be* never implies that it *is*. But at any rate, it's a start.

REFERENCES

Bouchard, T. J., Jr., Lykken, D. T., McGue, M., Segal, N. L., & Tellegen, A. (1990). Sources of human psychological differences. The Minnesota Study of Twins Reared Apart. *Science, 250*, 223–228.

Cattell, R. B. (1971). *Abilities: Their structure, growth, and action.* Boston: Houghton Mifflin.

Dawkins, R. (1982). *The extended phenotype.* San Francisco: Freeman.

Loehlin, J. C., Horn, J. M., & Willerman, L. (1989). Modeling IQ change: Evidence from the Texas Adoption Project. *Child Development, 60*, 993–1004.

Plomin, R., DeFries, J. C., & Loehlin, J. C. (1977). Genotype–environment interaction and correlation in the analysis of human behavior. *Psychological Bulletin, 84*, 309–322.

Scarr, S., & Weinberg, R. A. (1977). Intellectual similarities within families of both adopted and biological children. *Intelligence, 1*, 170–191.

Scarr, S., & Weinberg, R. A. (1978). The influence of "family background" on intellectual attainment. *American Sociological Review, 43*, 674–692.

Wechsler, D. (1981). *WAIS-R Manual.* New York: The Psychological Corporation.

Chapter 14

The Environmental Determinants of Intelligence Are Biological Factors Operating Prenatally and in Early Childhood

Richard Lynn

University of Ulster, Coleraine, Northern Ireland

I do not think that any of the contributors to this book managed to hit the nail on the head on the problem of environment effects on intelligence. None of them stated the four crucial points that I will first state summarily and then elaborate on in turn.

1. Environmental effects on intelligence are appreciable among children but decline among older adolescents and adults. Among adults, environmental factors account for only about 17% of the variance.

2. Contrary to the views of the contributors to this volume, the relevant environmental factors are not family influences affecting all children in a fam-

ily, such as parents' socioeconomic status, style of child rearing, quality of schooling they provide for their children, etc.

3. The environmental factors affecting intelligence are unique to individuals, for instance, one sibling in a family might suffer brain damage having an impairing effect on intelligence while the other siblings are unaffected.

4. These environmental influences must consist largely of biological and physical factors operating prenatally and in early infancy.

ENVIRONMENTAL EFFECTS DECLINE FROM CHILDHOOD TO ADULTHOOD

There is widespread agreement that the strength of environmental influences on intelligence declines while that of genetic influences increases from childhood to adulthood. Pedersen, Plomin, Nesselroade and McClearn (1992) estimated the heritabilities of intelligence to be about .50 for children and .80 for adults. Bouchard (1993) agreed and combined the data from numerous studies to give heritabilities of .42 for 4–6-year-olds, .53 for 6–16-year-olds, .57 for 16–20-year-olds and .75 for adults. These figures are calculated from the differences between identical and fraternal twins reared in the same families. Data from Monozygotic twins reared in different families produced similar results. The correlation between the pairs is a direct measure of heritability. Bouchard, Lykken, Tellegren, and McGue (1993) summarized the data consisting of 190 adult pairs yielding a correlation of .75, and Jensen (1993) reached the same conclusion. There is some difference of opinion about whether this figure should be corrected for test unreliability. Pederson et al. (1992) and Bouchard (1993) did not make this correction, but Jensen (1993) did. He assumed a test reliability of .90, and correction for this raises the heritability to .83. In my opinion, this is the correct procedure and gives an 83 percent heritability for intelligence among adults, leaving 17 percent for environmental effects.

There are two reasons for the diminishing influence of environmental factors during childhood and adolescence. Pederson et al. (1992) suggested that some of the genes determining intelligence may be "turned on" only at specific stages during the life span, analogous to the genes controlling the output of sex hormones being switched on at puberty. This may account for the jump in the heritability of intelligence from .42 among 4–6-year-olds to .53 among 6–16-year-olds, but it seems unlikely that hitherto latent genes for intelligence begin switching on after the age of 20 years to account for the jump in heritability from .57 among 16–20-year-olds to .75 among adults calculated by Bouchard (1993). Probably, the more important reason for these increasing heritabilities with age is that some environmental factors have temporary effects in childhood that fade out by adulthood.

It is environmental effects on the intelligence of adults that are the important ones. There is little interest and no advantage in producing an environmentally induced boost to intelligence in children, say, by the provision of nursery school education or whatever, if this vanishes after a few years and leaves no effect on the adult. Most of the environmental effects proposed by the contributors to this book are of this kind. My conclusion is that any consideration of the power of environmental effects on the intelligence of adults should begin by recognizing that these are quite small and that we are only dealing with about 17% of the variance.

THE RELEVANT ENVIRONMENTAL FACTORS ARE NOT FAMILY INFLUENCES AFFECTING ALL CHILDREN IN A FAMILY

The contributors to this book believe that the major environmental influences affecting intelligence are family effects such as styles of child rearing, cognitive stimulation, quality of schooling, and so on. This is not the case. The evidence leading to this conclusion is the zero correlation of the adult IQs of pairs of unrelated adopted children reared in the same family. There are three studies of these pairs of adopted children. The correlations obtained are –.03 (Scarr & Weinberg, 1978), .02 (Teasdale & Owen, 1984), and .02 (Willerman, 1987). Compiling the three studies gives a correlation of zero, which can only mean that differences in the way parents bring up their children have no effect on the child's intelligence once the child is adult. This rules out any causal effect of these child rearing practices and their correlates such as the parents' socioeconomic status, the attention and cognitive stimulation they give to their children, the books they provide, the kinds of school to which they send their children, and so on.

None of the contributors to this book seem to have realized the significance of this conclusion to be drawn from the zero correlation of the IQs of pairs of adopted children reared in the same family. Most of the contributors proposed environmental effects of precisely this kind such as "maternal affective tone," "parental expectation for achievement," "parental involvement," etc. listed by Wachs (Chapter 6) and "encouragement of exploration," "celebration of developmental advances," etc. listed by Ramey and Blair (Chapter 5). If these different styles of parenting had a permanent effect on the child's intelligence, there would be a positive correlation between the IQs of pairs of adopted children reared in the same family and tested as adults, but there isn't.

There is, however, a positive correlation for intelligence among unrelated pairs of adopted children reared in the same families and intelligence tested in childhood. Jensen (1993) calculated it for 570 pairs with an average age of 9 years at .29 or .32 corrected for attenuation. This shows that family styles of upbringing affects the IQ of children but the effects do not last into adulthood. The reason for this is that family experiences can provide temporary boosts for intelligence that subsequently evaporate. Family styles of upbringing are variously referred to in the literature as between-family variance, common environment, shared environment, and systematic environment. They all mean the same thing. They have no permanent effect on intelligence.

THE ENVIRONMENTAL FACTORS AFFECTING INTELLIGENCE ARE UNIQUE TO INDIVIDUALS

Environmental effects on intelligence consist of unique experiences, variously described in the literature as within-family variance, specific environment, nonshared environment, and random environment. The magnitude of these effects can be quantified as the residual after the variance due to genetic factors and between-family factors have been calculated. These are 83 percent and 0 percent, respectively, leaving 17 percent attributable to within-family variance. Alternatively, Jensen (1993) calculated, within-family variance as the reliability of the test minus the correlation for MZ twins reared apart. The two figures are .90 and .75 respectively, giving a figure of .15 for within-family environmental effects, virtually the same as the .17 calculated above.

ENVIRONMENTAL INFLUENCES CONSIST OF BIOLOGICAL AND PHYSICAL FACTORS

What are the nonshared, within-family environmental effects? It is proposed by Jensen (1993), in my opinion, correctly, that they must consist largely of a number of small biological and physical factors mainly operating prenatally or in early infancy. This is the inference to be drawn from the conclusion that the relevant environmental effects operate within families and are unique to one sibling and not to others. What kind of effects could these be? They will probably consist of such things as the quality of nutrition supplied by the mother to the foetus and traumas and illnesses of various kinds such as anoxia during childbirth causing some impairment to the brain. Some of these factors are detectable and well established in the research literature, although apparently unknown to the contributors to this book; for example, when monozygotic twins differ in birth weight due to differences in the nutrition supplied from the mother's placenta, the lighter twin at birth has permanently impaired intelligence that is detectable in adolescence and in adulthood. Seven studies showing these effects are summarized in Lynn (1990).

The conclusion that the environmental effects on intelligence consist largely of biological factors operating prenatally and in early childhood is confirmed by evidence on the secular increase in intelligence over the last 60 or so years. Flynn (Chapter 2) deals with this issue in the present volume, but he neglects to point out one of the most interesting features of this secular increase. This is that the increase, which amounts to approximately 3 IQ points per decade, is fully present among 5-year-olds and even among 2–3-year-olds on developmental tests (Flynn, 1984; Lynn, 1990). This shows that the factors responsible for the increase must be prenatal or operative before the age of 2–3 years. They cannot be anything to do with conjectural secular improvements in cognitive stimulation, education, TV, and so on, to which they are frequently ascribed. The fact that height and brain size have also shown secular increases of the same magnitude as intelligence, approximately one standard deviation over the last half century, confirms the inference that the cause lies in biological factors consisting principaly of improved nutrition. I have set out the evidence and arguments for this conclusion in Lynn (1990).

REFERENCES

Bouchard, T. J. (1993). The genetic architecture of human intelligence. In P. A. Vernon (Ed.), *Biological approaches to the study of human intelligence* (pp. 33–94). Norwood, NJ: Ablex Publishing Corp.

Bouchard, T. J., Lykken, D. T., Tellegren, A., & McGue, M. (1993). Genes, drives, environment and experience: EPD theory—Revised. In C. P. Benbow & D. Lubinski (Eds.), *From psychometrics to giftedness: Essays in honor of Julian C. Stanley* (pp. 106–135). New York: Academic Press.

Flynn, J. R. (1984). The mean IQ of Americans. *Psychological Bulletin, 95*, 29–51.

Jensen, A. R. (in press). The puzzle of nongenetic variance. In R. J. Sternberg & E. L. Grigorenko (Eds.), *Intelligence, Heredity and Environment.* New York: Academic Press.

Lynn, R. (1990). The role of nutrition in the secular increases in intelligence. *Personality and Individual Differences, 11*, 273–285.

Pedersen, N. L., Plomin, R., Nesselroade, J. R., & McClearn, G. E. (1992). A quantitative genetic analysis of cognitive abilities during the second half of the life span. *Psychological Science, 3*, 346–353.

Scarr, S., & Weinberg, R. A. (1978). The influence of family background on intellectual attainment. *American Sociological Review, 43*, 674–692.

Teasdale, T. W., & Owen, D. R. (1984). Heredity and familial environment in intelligence and education level—A sibling study. *Nature, 309*, 620–622.

Willerman, L. (1987, July). Where are the shared environmental influences on intelligence and personality? Paper presented at the Society for Research in Child Development, Baltimore, MD.

Chapter 15

On the Role of the Physical Environment in the Development of Intelligence

Ernesto Pollitt

Department of Pediatrics and Program of International Nutrition, University of California, Davis, California

Carmen Saco-Pollitt

Department of Teacher Education, California State University, Sacramento, California

As psychologists move towards the construction of a theory of human intelligence, it is increasingly clear that there is a need for fine-grain analyses of the role that distinct aspects of context play in shaping its development (Ceci, 1993). There is a need to go beyond the context of highly industrialized, high-income societies, particularly in regards to the physical environment. Advanced technology, economic growth, and public health policies have led to the control of aspects of the physical environment (e.g., food security) that still determine, in part, the survival and continuity of families in low- and middle-income countries. Some of these physical contextual factors are likely to have direct effects on brain growth and possibly on the development of intelligence. As noted below, millions of people across different world regions live in such physical environmental contexts.

We will argue that distinct factors in the natural environment contribute directly to the variability observed in human development. We make no attempt for a comprehensive review of the available evidence; our effort is restricted to making a case for the need to adopt a broader view in the process of theory construction on the development of human intelligence. In our view, extreme environments test the limits of human adaptability and, as such, are an obligatory forum for testing theories of human development which claim universality.

In particular, our argument is based on evidence that suggests that iodine deficiency in different regions throughout the continents, low oxygen pressure in the high altitudes (e.g., the Andes, Himalayas), and intestinal parasites (e.g., geohelminths) in the tropics and subtropics influence the development of the central nervous system (CNS). Such effects are analogous to the effects that these factors have on vital functions and human growth.

We do not judge the success of adaptation or of intellectual competence of the societies under consideration, nor are we concerned with particular cognitive and behavioral outcomes in adulthood. We concern ourselves with direct effects on the development of the CNS, and with the results of tests of intellectual function in children. By direct effects, we mean changes secondary to, or activated by, biological mechanisms without an apparent or obvious participation of cultural or social factors.

IODINE

Iodine is a trace element necessary for human growth and development available in the sea, air, and land. While it was present during the primordial development of the earth, over time, rain and snow leached it away from the soil and moved it to the sea. This effect of rain on iodine is also the main determinant of iodine-deficient regions (Stanbury & Hetzel, 1980), particularly in mountainous areas with abundant precipitation.

At one time, there had been a high prevalence of goiter in the Great Lakes region of the United States. In 1924 the prevalence of goiter in four Michigan counties ranged from 26.0 to 64.4 percent. The iodization of salt and changes in food technology that have led to a high intake of iodine throughout the country have led to a negligible prevalence of iodine deficiency today in the United States (Matovinovic & Trowbridge, 1980). However, about 100 million people in the world are still at risk for iodine deficiency (United Nations Subcommittee on Nutrition, 1992).

Iodine is an essential component of thyroxine (T_4) and triiodothyronine (T_3) hormones, which are needed for normal growth and development. Iodine deficiency has been associated with increased perinatal mortality, stillbirths, low birth weight, and increased morbidity (Levin, Pollitt, Galloway, & McGuire, 1993). Deficiencies in maternal and fetal thyroid hormone level have severe, multiple effects on brain structural growth and biochemical maturation (Hetzel, Mano, & Chevadev, 1988; Porterfield & Hendrich, 1993). These effects are also observed as a result of early postnatal thyroid deficiencies. While their nature and severity vary as a function of the timing of the insult, early fetal hormonal dysfunction is associated with retarded cerebral neurogenesis and cell migration and impaired neuronal differentiation (Boyages & Halpern, 1993).

Cretinism, defined as mental deficiency coupled with defects of hearing and speech, squint, and disorders of stance and gait, is most frequently caused by severe iodine deficiency (Boyages & Halpern, 1993). This condition, however, is not our direct concern at this time. Of potentially greater public health significance is the association between mild iodine deficiency and a modest degree of impairment of the CNS (Bautista, Barker, Dunn, Sanchez, & Kaiser, 1982; Bleichrodt, Drenth, & Querido, 1980; Pharoah, Connolly, Ekins, & Harding, 1984). While the evidence on causality in human populations is inconclusive in light of shortcomings in design, it is certainly suggestive, particularly in light of the supporting evidence from animal experiments (DeNayer & Dozin, 1988). Ethical considerations regarding the use of placebo for at-risk subjects have prevented the use of randomized clinical trials to test the main hypothesis that mild iodine deficiency impairs the CNS.

Illustrative is a study conducted by Boyages et al. (1989) of inhabitants of two rural villages in north-central China. These villages did not differ from each other in geographical remoteness, population size, social class distribution, or types of schools. There was a significant difference in the degree of risk for iodine deficiency, however. In the high-risk village, the school children's performance on the Hickey Nebraska Test of Learning Aptitude was significantly poorer than that of similar-aged children in the low-risk village. The children in the first village also showed a higher prevalence of abnormal neurological signs than those in the second village.

HIGH ALTITUDE

Bolivia, China, Ethiopia, India, Nepal, and Peru are among the countries with populations living above an altitude of 3,000 meters (10,000 feet.) and are exposed to a biophysical environment distinct from other places in the world. In 1970, it was estimated that these populations totaled 13 to 16 million people (De Jong, 1970). The high-altitude (HA) environment includes low oxygen and barometric pressure, reduced relative humidity, reduced gravitational force, increased solar radiation, and marked variations of temperature in short periods (Thomas & Winterhalder, 1976).

While there are differences in cold, humidity, and radiation between HA regions independent of latitude, barometric and oxygen pressure remain constant at the same altitude across the world. Reduced O_2 has received, by far, the greatest emphasis in the biomedical literature on the effects of HA on human populations (Baker, 1978; Baker & Little, 1976; Heath & Williams, 1989; Winslow & Monge, 1987). This is not surprising in light of the established adverse effects of reduced oxygen on human physiology.

In HA areas, there is an increased prevalence of intrauterine growth retardation and low birth weight (McClung, 1969). The chances of neonatal and infant survival are also greater in HA areas than in lower-altitude areas (Baker, 1984). These effects are observed not only in low-income rural populations in HA regions of developing countries, but also in the high-altitude towns and cities of Colorado in the United States (Yip, 1989).

The growth and maturation of the brain in rodents are adversely affected by exposure to HA conditions that simulate those found among human populations living at about 3,500 to 4,000 meters (Cheek, Graystone, & Rowe, 1969; Petropoulos, Vernadakis, & Timiras, 1969). For example, cell multiplication, content, and myelogenesis are comparatively slow in rat pups exposed to HA conditions, and these delays covary with functional CNS impairments such as immature response to electroshock treatment (Timiras & Wooley, 1966).

After controlling for ethnic background, 36- and 48-hour-old infants born at about 4,200 meters presented less fully developed interactive and motoric abilities as compared with infants born at sea level (Saco-Pollitt, 1989). The infants born at HA were less likely to be attentive to environmental stimuli in general and less likely to be oriented to purposely presented visual and auditory stimuli. A similar directional difference was observed in the degree of the infants' responsiveness to being held by the examiner. In comparison with the infants in low-altitude regions, those at high altitudes were less cuddly. The neonates at the HA were also less active and more hypotonic; their movements were more "jerky" (or less smooth), and they were less likely to keep their heads up when they were moved from a prone to a sitting position. In addition, the infants at low altitude were more successful than those at HA in quieting themselves.

Further, the comparison of 2–12-month-old infants from two genetically similar rural groups at high and low altitudes showed a statistically significant difference ($p = 0.01$) in motor development in favor of the low-altitude sample (Haas et al., 1982). These differences were not present, however, among 13–27-month-old children from the same groups.

Symptoms of malaise (e.g., headache, lethargy, vomiting) and discomfort may subside 24–48 hours after exposure to high altitude conditions. However, selective impairments of CNS function are likely to continue in the sojourner for much longer periods. There was a marked decline and a slow recovery in eye–hand coordination tests among young adults with HA exposure ranging from 1–25 months (Sharma, Malhotra, & Baskaran, 1975). Rather than making a rapid adjustment to HA conditions, these subjects took over 10 months to show signs of CNS recovery.

INTESTINAL PARASITES

Over two billion people carry intestinal parasites including *Ascaris lumbricoides, Trichuris trichiura*, and hookworm (Stephenson, 1987). These worms often coexist in the human host and are more likely to be found in the preindustrial regions of the tropics and subtropics. For the purposes of this chapter, the focus is restricted to hookworm because its effect on CNS function through a depletion of iron in the organism is clearer than that of the other species. Prevalence of hookworms in communities at risk increases linearly with age, peaking in adolescence and young adulthood. Thereafter, it stabilizes or drops gradually (Bundy, 1990; Pawlowski, Schad & Stott, 1991).

The larva of the hookworm penetrates the human host through the skin (usually between the toes, in the legs, and the buttocks), enters the circulation of the body, and is transported to the heart and lungs. Once it crosses the alveoli and migrates to the bronchiolus, bronchi, and trachea, it enters the pharynx. After it is swallowed, the larva moves through the stomach to the small intestine where it will mature, feeding on the intestinal mucosa and blood. Six to eight weeks later a mature hookworm will lay eggs that are expelled from the host with the feces (Latham, 1990; Pawlowski, Schad, & Stott, 1991; Stephenson, 1987).

On each site of the intestine where the hookworm feeds, bleeding occurs; high intensity of infection can cause iron deficiency anemia (Crompton & Stephenson, 1990; Layrisse & Roche, 1964; Misagena, Gilles, & Maebraith, 1972; Stephenson, 1987). A person would have to double the recommended iron intake (10 mg/d for males and 18 mg/d for females of child-bearing age) to maintain replete iron stores (Stephenson, 1987).

Fetal growth is at risk among pregnant women with severe anemia (Brabin, 1988; Lieberman, Ryan, Monson, & Schoenbaum, 1987, 1988). Maternal mortality is also increased (Fleming, 1989) and the capacity of the organism to resist infections is compromised (Chandra & Puri, 1985).

In the early 1900s, when intestinal parasites were still a public health problem in the United States, carriers of hookworms in school lagged behind those free of infection (Kelley, 1917; Smillie & Spencer, 1926; Stiles, 1915a, 1915b; Strong, 1916). Similarly, during the First World War, new army recruits infected with hookworm were physically and mentally less capable than noncarriers. As would be expected, these limitations were also observed among men involved in manual labor in the tropics.

There is a positive association between levels of body iron and performance scores on mental and motor development scales in infants and toddlers and cognitive tests in older children (Lozoff, 1990; Pollitt, 1993). However, iron repletion produces significant improvement in subjects (Aukett, Parks, Scott, & Wharton, 1986; Idjradinata & Pollitt, 1993). Furthermore, the size of the pretreatment differences in test performance observed between subjects with and without anemia was similar to the pre- to posttreatment changes observed in the anemic subjects.

Because of the compensatory mechanisms available to the human organism, the cause of the effects is not likely to be a deficit in oxygen transport. It is hypothesized, based on seminal studies, that a drop in the availability of cerebral iron is the culprit, even at the early stages of iron deficiency. D2 receptors of dopamine in the brain are altered in the presence of iron deficiency in rodents (Youdim, 1990).

FINAL NOTE

The adverse effects we have discussed may benefit population survival and continuity. While there are no data available to support such contention, this line of reasoning agrees with recent arguments on the functional significance of anemia (Kent, 1992; Stuart-Macadam, 1992). The removal of circulating iron reduces its availability to infective pathogens and decreases the probabilities of infection (see Stuart-Macadam & Kent, 1992). This effect is adaptive under conditions where infant and child mortality is primarily determined by gastrointestinal (diarrhea) and respiratory infections.

A judgement on competence calls for assessments of the different functions in which a particular cognitive process is involved. While tests of cognition may show, for example, differences in working memory between carriers and non-carriers of hookworm, such a difference is insufficient to claim a lower level of competence in those that are infected. This potential alteration might contribute to the operation of other cognitive components that are particularly suitable for tropical conditions, where helminthiasis is endemic. A judgement on adaptedness must await a full assessment of all the attributes of the cognitive function. Cultural practices may also act as a buffer against physical conditions that operate as environmental stressors on the organism.

The degree of human stress associated with the physical variables unique to those environments which house millions of people in the developing world makes us wonder about their role in the development of intelligence of the populations under scrutiny. In line with cross-cultural researchers who emphasize the different cultural features of the environment in explanations of human development, we believe that the exclusion of the physical and biological factors deserve such attention.

REFERENCES

Aukett, M. A., Parks, Y. A., Scott, P. H., & Wharton, B. A. (1986). Treatment with iron increases weight gain and psychomotor development. *Archives of Disease in Childhood, 61*, 849–857.

Baker, P. T. (1978). *The biology of high altitude people*. Cambridge: Cambridge University Press.

Baker, P. T. (1984). The adaptive limits of human populations. *Man, 19*, 1–14.

Baker, P. T., & Little, M. A. (Eds.). (1976). *Man in the Andes: A multidisciplinary study of high-altitude Quechua*. Stroudsburg, PA: Dowden, Hutchinson & Ross.

Bautista, A., Barker, P. A., Dunn, J. T., Sanchez, M., & Kaiser, D. L. (1982). The effects of oral iodized oil on intelligence, thyroid status, and somatic growth in school-age children from an area of endemic goiter. *American Journal of Clinical Nutrition, 35*, 127–134.

Bleichrodt, N., Drenth, P. J. D., & Querido, A. (1980). Effects of iodine deficiency on mental and psychomotor abilities. *American Journal of Physical Anthropology, 53*, 55–67.

Boyages, S. C., Collins, J. K., Maberly, G. F., Jupp, J. J., Morris, J., & Eastman, C. J. (1989). Iodine deficiency impairs intellectual and neuromotor development in apparently normal persons. *The Medical Journal of Australia, 15*, 137–141.

Boyages, S. C., & Halpern, J. P. (1993). Endemic cretinism: Toward a unifying hypothesis. *Thyroid, 3*, 59–69.

Brabin, B. J., Ginny, M., Sapau, J., Galme, K., & Paino, J. (1990). Consequences of maternal anemia on outcome of pregnancy in a malaria endemic area in Papua New Guinea. *Annals of Tropical Medicine and Parasitology, 84*(1), 11–24.

Bundy, D. A. P. (1990). Is the hookworm just another geohelminth? In G. A. Schad & K. S. Warren (Eds.), *Hookworm disease: Current status and new directions* (pp. 147–164). London: Taylor and Francis.

Ceci, S. J. (1993). Contextual trends in intellectual development. *Developmental Review, 13*, 403–435.

Chandra, R. K., & Puri, S. (1985). Trace element modulation of immune responses and susceptibility to infection. In R. J. Chandra (Ed.), *Trace elements in nutrition and children* (pp. 87–106). Nestle Nutrition Workshop Series, Vol. 8. New York: Raven Press.

Cheek, D. B., Graystone, J. E., & Rowe, R. D. (1969). Hypoxia and malnutrition in newborn rats: Effects on RNA, DNA, and protein in tissues. *American Journal of Physiology, 217*, 642–645.

Crompton, D. W. T., & Stephenson, L. S. (1990). Hookworm infection, nutritional status, and productivity. In G. A. Schad & K. S. Warren (Eds.), *Hookworm disease: Current status and new directions* (pp. 231–264). Engelwood Cliffs, NJ: Prentice-Hall.

De Jong, G. F. (1970). Religious fundamentalism, socioeconomic status, and fertility attitudes in the Southern Appalachians. In T. R. Ford & G. F. De Jong (Eds.), *Social demography* (pp. 226–235). Englewood Cliffs, NJ: Prentice-Hall.

DeNayer, P., & Dozin, B. (1988). Thyroid hormone receptors in the developing brain. In G. R. DeLong, J. Robbins, & P. G. Condliffe (Eds.), *Iodine and the brain* (pp. 51–58). New York: Plenum Press.

Fleming, A. F. (1989). *Consequences of anaemia in pregnancy on the mothers.* Unpublished manuscript, Liverpool School of Tropical Medicine, Department of Tropical Medicine and Infectious Diseases, Liverpool.

Haas, J. D., Moreno-Black, G., Frongillo, E. A., Jr., Pabon, J., Pareja, G., Ybarnegaray, J., & Hurtado, L. (1982). Altitude and infant growth in Bolivia: A longitudinal study. *American Journal of Physical Anthropology, 59*, 251–262.

Heath, D., & Williams, D. R. (1989). *High altitude medicine and pathology.* London: Butterworths.

Hetzel, B. S., Mano, M. L., & Chevadev, J. (1988). The fetus and iodine deficiency: Marmoset and sheep. In G. R. DeLong, J. Robbins, & P. G. Condliffe (Eds.), *Iodine and the brain* (pp. 177–186). New York: Plenum Press.

Idjradinata, P., & Pollitt, E. (1993). Reversal of developmental delays in iron-deficient anaemic infants treated with iron. *The Lancet, 341*, 1–4.

Kelley, T. L. (1917). The effect of malaria and hookworm upon physical and mental development of school children. *Elementary School Journal, 18*, 43–51.

Kent, S. (1992). Anemia through the ages: Changing perspectives and their implications. In P. Stuart-Macadam & S. Kent (Eds.), *Diet, demography, and disease: Changing perspectives on anemia* (pp. 1–30). Hawthorne, NY: Aldine De Gruyter.

Latham, M. C. (1990). Ascariasis and hookworm disease—Their impact on human nutrition. *Postgraduate Doctor: Middle East, 13*, 118–127.

Layrisse, M., & Roche, M. (1964). The relationship between anemica and hookworm infection. *American Journal of Hygiene, 79*, 279–301.

Levin H. M., Pollitt, E., Galloway, R., & McGuire J. (1993). Micronutrient deficiency disorders. In D. T. Jamison, W. H. Mosley, A. R. Measham, & J. L. Bobadilla (Eds.), *Disease control priorities in the developing world* (pp. 421–451). New York: Oxford Medical Publications.

Lieberman, A., Ryan, K. J., Monson, R. R., & Schoenbaum, S. C. (1987). Risk factors for racial differences in the rate of premature births. *New England Journal of Medicine, 317*, 743–748.

Lieberman, A., Ryan, K. J., Monson, R. R., & Schoenbaum, S. C. (1988). Association of maternal hematocrit with premature labor. *American Journal of Obstetrics and Gynecology, 159*, 107–114.

Lozoff, B. (1990). Has iron deficiency been shown to cause altered behavior in infants? In J. Dobbing (Ed.), *Brain, behaviour, and iron in the infant diet* (pp. 107–131). London: Springer–Verlag.

McClung, J. (1969). *Effects of high altitude on human birth.* Cambridge, MA: Harvard University Press.

Matovinovic, J., & Trowbridge, F. L. (1980). North America. In J. B. Stanbury & B. S. Hetzel (Eds.), *Endemic goiter and endemic cretinism: Iodine nutrition in health and disease* (pp. 31–67). New York: John Wiley and Sons.

Misagena, S., Gilles, H. M., & Maebraith, B. G. (1972). Studies in *Ancylostoma caninum* infection in dogs. II: Anatomical changes in the gastrointestinal tract. *Annals of Tropical Medicine and Parasitology, 66*, 203–207.

Pawlowski, Z. S., Schad, G. A., & Stott, G. J. (1991). *Hookworm infection and anaemia: Approaches to prevention and control.* Geneva: World Health Organization.

Petropoulos, E. A., Vernadakis, A., & Timiras, P. S. (1969). Nucleic acid content in developing rat brain after prenatal and/or neonatal exposure to high altitude. *Federation Proceedings, 28*, 1001–1005.

Pharaoh, P. O. D., Connolly, K. J., Ekins, R. P., & Harding, A. G. (1984). Maternal thyroid hormone levels in pregnancy and the subsequent cognitive and motor performance of the children. *Clinical Endocrinology, 21*, 265–270.

Pollitt, E. (1993). Iron deficiency and cognitive function. *Annual Review of Nutrition, 13*, 521–537.

Porterfield, S. P., & Hendrich, C. W. E. (1993). The role of thyroid hormones in prenatal and neonatal neurological development—Current perspectives. *Endocrine reviews, 14*, 94–106.

Saco-Pollitt, C. (1989). Ecocultural context and developmental risk: Birth in the high altitudes (Peru). In J. K. Nugget, B. M. Lester, & T. B. Brazelton (Eds.), *The cultural context of infancy, Vol. 1: Biology, culture, and infant development* (pp. 3–25). Norwood, NJ: Ablex Publishing Corp.

Sharma, V. M., Malhotra, M. S., & Baskaran, A. S. (1975). Variations in psychomotor efficiency during prolonged stay at high altitude. *Ergonomics, 18*(5), 511–516.

Smillie, W. G., & Spencer, R. (1926). Mental retardation in school children infested with hookworm. *Journal of Educational Psychology, 17*, 314–321.

Stanbury, J. B., & Hetzel, B. S. (Eds.). (1980). *Endemic goiter and endemic cretinism: Iodine nutrition in health and disease.* New York: John Wiley and Sons.

Stephenson, L. S. (1987). *The impact of helminth infections on human nutrition.* London: Taylor & Francis.

Stiles, C. W. (1915a). The school grades attained by 2,166 White school children in the city of X, classified by age, sanitation, and intestinal parasites. *U.S. Public Health System Report, 30.*

Stiles, C. W. (1915b). The school grades attained by 2,166 White school children (1,062 boys, 1,104 girls) in the city of X, classified by age, sanitation, and intestinal parasites. *U.S. Public Health System Report, 21.*

Strong, E. K. (1916). *Effects of hookworm disease on the mental and physical development of children.* New York: The Rockefeller Foundation. (International Health Commission; Publication No. 3)

Stuart-Macadam, P. (1992). Anemia in past human populations. In P. Stuart-Macadam & S. Kent (Eds.), *Diet, demography, and disease: Changing perspectives on anemia* (pp. 151–170). Hawthorne, NY: Aldine De Gruyter.

Stuart-Macadam, P., & Kent, S. (Eds.). (1992). *Diet, demography and disease: Changing perspectives on anemia.* Hawthorne, NY: Aldine de Gruvter.

Thomas, R. B., & Winterhalder, B. P. (1976). Physical and biotic environment of southern highland Peru. In P. T. Baker & M. A. Little (Eds.), *Man in the Andes: A multidisciplinary study of high-altitude Quechua* (pp. 21–59). Stroudsburg, PA: Dowden, Hutchison & Ross.

Timiras, P. S., & Wooley, D. E. (1966). Functional and morphological development of brain and other organs at high altitude. *Federation Proceedings, 25*, 1312–1320.

United Nations Subcommittee on Nutrition. (October 1992). Second report on the world nutrition situation, Vol I.

Winslow, R. M., & Monge, C. (1987). *Hypoxia, polycythemia, and chronic mountain sickness.* Baltimore: Johns Hopkins University Press.

Yip, R. (1989). The interaction of lead and iron. In L. J. Filer, Jr. (Ed.), *Dietary iron: Birth to two years* (pp. 179–181). New York: Raven Press.

Youdim, M. B. H. (1990). Neuropharmacological and neurobiochemical aspects of iron deficiency. In J. Dobbing (Ed.), *Brain, behaviour, and iron in the infant diet* (pp. 83–106). London: Springer–Verlag.

Chapter 16

Commentary on the Contributions to this Volume

Herman H. Spitz

Princeton, NJ

It goes without saying, although it has been said ad nauseam, that both heredity and environment contribute to the development of the maturing individual, including the expression of intelligence (see, for example, the recently published *Nature, Nurture, & Psychology*, edited by Plomin and McClearn, 1993). The fact that there is general agreement on this point would lead one to expect an exciting and revealing decade ahead. But just when things seem to be going along swimmingly there is this little problem waiting to gum up the works. "This intelligence you people have been measuring," goes the argument, "is not really intelligence at all. In fact, there is no such thing as general intelligence (*g*), as Spearman would have it, only many separate aptitudes or, as they used to say, faculties." Furthermore, they say, one need not even agree with Gardner's (1983) list of intelligences to claim that what IQ tests have been measuring is only academic intelligence, leaving much else unmeasured (Sternberg, 1985).

If this argument has merit, then it follows that the influence of nature and nurture will have to be assessed for each of the separate intelligences. Of course, there are those (e.g., Jensen, 1986) who do not believe that we should dispense with *g*, and consequently the multiple/general debate is heating up just as the nature/nurture debate appears to be heading toward rapprochement.

Nested in the general nature/nurture debate is the attempt to estimate the approximate variance in intelligence (typically measured by intelligence tests) due to heredity compared to the variance contributed by the environment. Although this issue has benefited greatly from the work on identical twins reared apart (Bouchard, Lykken, McGue, Segal, & Tellegen, 1990) and the creative methods of behavioral geneticists (Plomin, DeFries, & McClearn, 1990), it is of more theoretical than practical relevance because even small environmental variance allows for very large environmental effects. However, the statement that heredity is not destiny is usually followed by examples in the physical realm, such as the increased height resulting from the administration of growth hormones. Evidence that the genes do not have an implacable influence on intelligence, on the other hand, has been provided more often by environmental restrictions on the growth of intelligence than by demonstrations that pedagogical or psychological interventions can raise its level (Spitz, 1986).

Researchers interested in hereditary sources of variance have an advantage: they can say that genes are responsible for a certain kind of behavior, but of course they don't know *which* genes. Environmentalists, on the other hand, know that environmental sources of variance are out there somewhere, and have the unenviable task of trying to find them; unenviable because individuals are bombarded by environmental stimuli and the researchers cannot follow subjects around every minute, nor can they always know to which stimuli individuals are attending, nor how they internally respond. They cannot be privy to dreams and the workings of the unconscious (Spitz, 1993).

Although these are my choices for the paramount issues in current intelligence research, others have their own favorites. For example, even if we resolve the multiple/general problem, we still have to define the *process* that occurs when one individual performs more intelligently than another. There is, too, the perplexing rise in IQ over generations (see Flynn, this volume). And some issues are clearly bigger, in a scientific sense, than my selection: the question of how intelligence evolved, for example, and the looming presence of molecular biology and the search for intelligence genes.

How do the environmentally oriented contributors to the present volume deal with what I have suggested are the paramount issues? There are no Kaminites here; no one believes that genes have nothing to do with the variance in intelligence (Kamin, 1974) and no one claims that heredity and environment do not interact. None are much interested in the question of multiple intelligences versus *g*. With the exception of Borkowski and Dukewich, they do not estimate the proportion of variance attributable to certain environmental variables. However, the one thing many of them are willing to do (with varying degrees of certainty) is to suggest sources of environmental variance.

Consequently we can build a cumulative list of environmental variables that the contributors believe influence the development of intelligence.

Borkowski and Dukewich "speculate that 15 to 20 percent of the IQ variance in children, ages 10–14, might be due to differences in self-regulatory skills." They suggest that good self-regulatory skills are induced primarily by secure attachment to a caregiver, which in turn is influenced by the stability of the neighborhood and the quality of social supports.

Flynn is searching for environmental variables that raise scores on *tests* of intelligence (IQ) without affecting *actual* intelligence. In passing, however, he speculates that personal traits "like a person's work ethic, investment of self-esteem in academic and occupational achievement, self-discipline, and sobriety look more important [in producing high achievement by Chinese-Americans and raised IQs following early intervention] than middle-class affluence, educational toys, and stimulating leisure-time activities." He does not name the environmental variables that might influence the development of these personal traits.

Keating touches on all the issues I have mentioned and many more, including the domain-specificity versus domain-generality debate, which differs from the question of multiple intelligences versus general intelligence. Without naming the environmental sources that influence (interact with) them, he mentions a number of personal variables, including automaticity, procedural and declarative knowledge, "self-regulating cognitive activities (like metacognitive strategies or control processes); and social, emotional, and motivational factors, all of which influence the development of intelligence."

Pianta and O'Connor believe that "multi-level, multi-influence" early intervention projects affect IQ scores (although I know of no evidence that the effect is permanent) "in part because they involve multiple components of the developmental niche: social and physical settings, parental beliefs about child-rearing, nutrition, medical care, or parent education/job training." They eschew the search for specific (narrow) environmental variables, but we can add these broad environmental variables to our list, acknowledging nevertheless that Pianta and O'Connor think of the environment as a *system* (see also Wachs, this volume) that is integrated with genetic and biological influences.

Ramey and Blair present a working model of intelligence that "has undergirded the conceptualizations of several early intervention programs." They offer six essential mechanisms for normal cognitive development: adults should help and encourage children to explore, teach them basic cognitive skills, reinforce their accomplishments, help them consolidate and extend new skills, protect them from "inappropriate disapproval, teasing or punishment," and provide them with a "a rich and responsive language environment." They propose four "carrier mechanisms" that maintain the beneficial results of these adult-mediated mechanisms, but I will not add them to our list.

Wachs gives us some additional candidates, including nutrition, quality of caregiving, cultural beliefs, school attendance, school or parental expectancies, the "availability, variety and responsivity of social and object stimulation," parental involvement, guidance by caregivers in the child's problem solving, and "lower levels of noise, crowding or restriction of exploration."

Wagner and Oliver make the relevant point that, "Answers to the questions of how, and how much is intelligence influenced by the environment are likely to be different for different conceptualizations of intelligence and environments." They are very sensitive to the difficulty in choosing environmental variables that are likely to influence the growth of intelligence. Nevertheless, noting the similarity with the development of exceptional performance, they implicate (gingerly) the long-term practice and teaching that is associated with attending school.

Although the primary importance of covariances and interactions is stressed, most of the contributors supply enough environmental variables (of varying breadth) for us to compile a sizable list. Where personal variables—such as procedural knowledge, motivation, or self-esteem—are mentioned without the environmental influences that may affect them, they are excluded (although there is no reason why a list of personal variables could not also be compiled). Here is the final list of environmental variables that may influence the development of intelligence:

access to enhancing experiences
adult-assisted consolidation and extension of new skills
adult-assisted exploration
adult-assisted teaching of basic skills
adult reinforcement
caregiver-guided child participation in problem solving
cultural beliefs
enhanced knowledge base
less restricted exploration
medical care
more accomplished mentors
nutrition
parental involvement
parental beliefs about childrearing
parent education/job training
quality of caregiving
quality of social supports
reduced noise and crowding
rich, responsive language environment
secure attachment to a caregiver (which induces self-regulatory skills)
social and physical settings
school attendance, and long-term practice associated with attending school

school or parental achievement expectancy
social and object stimulation
stability of neighborhoods

This is a small sample, taken from a single book; a search of the literature would increase it immeasurably. To repeat, the contributors do not intend to imply that these variables work in isolation without interacting (and covarying) with other variables, both genetic and environmental. Nevertheless, I believe at least some of the contributors would agree with me that the question that must be asked is this: What environmental variables do we select for interaction with genetic variables, and how can we advance as a science simply by introducing an endless series of variables of unknown relative importance to an already unmanageable list?

REFERENCES

Bouchard, T. J., Lykken, D. T., McGue, M., Segal, N. L., & Tellegen, A. (1990). Sources of human psychological differences: The Minnesota Study of Twins Reared Apart. *Science, 250*, 223–228.

Gardner, H. (1983). *Frames of mind: The theory of multiple intelligences.* New York: Basic Books.

Jensen, A. R. (1986). *g*: artifact or reality? *Journal of Vocational Behavior, 29*, 301–333.

Kamin, L. J. (1974). *The science and politics of IQ.* Hillsdale, NJ: Erlbaum.

Plomin, R., DeFries, J. C., & McClearn, G. E. (1990). *Behavioral genetics: A primer* (2nd ed.). New York: Freeman.

Plomin, R., & McClearn, G. E. (Eds.). (1993). *Nature, nurture, & psychology.* Washington, DC: American Psychological Association.

Spitz, H. H. (1986). *The raising of intelligence: A selected history of attempts to raise retarded intelligence.* Hillsdale, NJ: Erlbaum.

Spitz, H. H. (1993). The role of the unconscious in thinking and problem solving. *Educational Psychology, 13*, 229–244.

Sternberg, R. (1985). *Beyond IQ: A triarchic theory of human intelligence.* New York: Cambridge University Press.

Chapter 17

Where Are the Environmental Influences on IQ?

Lee Anne Thompson

Department of Psychology, Case Western Reserve University

I have just read seven eloquent and intricate chapters, each with a unique and thought-provoking perspective on how the environment *ought* to affect intellectual function. However, I'm still left with the burning question of "How does the environment really affect IQ? Why has the identification of environmental influences on intelligence been so elusive"?

I have structured my response to these chapters from the perspective that I feel most comfortable, that of a developmental behavioral geneticist. In other words, I limit myself to a discussion of the etiology of *individual differences* in intelligence as defined by variance in IQ scores.

With respect to IQ, the nature–nurture controversy is dead and should be permanently laid to rest. There is no longer sufficient reason to conduct a study of IQ solely to test whether genetic influences are operating (Bouchard, 1993). Clearly, genes influence IQ; however, acknowledging genetic contributions does not contradict the importance of environmental research. Additive genetic influences account for 50 to 70 percent of the variance in IQ, leaving an appreciable amount unexplained (Chipuer, Rovine, & Plomin, 1990). Common misconceptions about how genes affect behavior can interfere with acceptance of genetic influences. First, as pointed out by Wachs (Chapter 6), although genetic influences are most certainly biologically mediated, biological aspects of behavior can be influenced by the environment. Separating the environment from biology can only be counterproductive. Second, genetic influence does not negate environmental influence. Heritability is a population parameter. If the environment is completely homogeneous across a sample of children and does not differentially affect their IQ, then heritability will be high and environmental influences will be low or negligible. However, if "new" environmental variation is introduced that creates individual differences in IQ, then heritability will decrease and environmental parameters will become significant.

I would like to argue that behavioral genetic methodology provides one of the most informative approaches for the study of the environment. While I could discuss in depth the many contributions that behavioral genetic methodology has to offer the study of the environment in general, I will instead focus on three main findings of particular relevance to the study of IQ.

SHARED VS. NONSHARED ENVIRONMENT

Behavioral geneticists have used three broad categories of influences on behavior to structure their research designs: genetic, shared environment, and nonshared environment. Briefly, genetic influences refer to heritable variation, shared environment refers to those aspects of the environment that make family members (or members of a twin pair) similar to each other, and nonshared environment refers to those aspects of the environment that create dissimilarity among family members. While Wachs seems to feel that the distinction between shared and nonshared environmental influences is superfluous and heuristically uninformative, I would like to argue otherwise. A strong message has been sent via behavioral genetic studies that have used these broad classifications of the environment. Those aspects of the environment that have traditionally been thought to influence IQ, such as SES, parental occupation, and the number of books in the home, fall into the category of shared environment. In general, they have been found to be of trivial importance with respect to individual differences in IQ independent of genetics.

Although adoption studies find some adoptive parent/adopted child resemblance when the offspring are young, this resemblance disappears once the children become adolescents (Scarr, 1992). Recent studies of twins reared apart (Bouchard, Lykken, McGue, Segal, & Tellegen, 1990; Pedersen, Plomin, Nesselroade, & McClearn, 1992) provide the strongest evidence supporting this point; identical twins reared apart are as similar regarding IQ as are identical twins reared together. Nonshared environment appears to be the primary source of environmental variance affecting IQ.

The important question resulting from behavioral genetic research is, then, "What nonshared environmental influences affect IQ?" Behavioral genetic studies that attempt to identify systematic, nonshared environmental influences have not been very successful. An obvious candidate, birth order, appears to account for a miniscule amount of the variance in IQ. To the extent that the relevant nonshared environmental influences operate in an unsystematic, idiosyncratic fashion, nonshared environment will be a difficult area of study. A recent report (Molenaar, Boomsma, & Dolan, 1993) suggests that many studies have implicated the importance of nonshared environment because a large part of the variance is composed of nonlinear epigenetic effects which occur randomly. In addition, Jensen (in press) has suggested that like polygenic effects on IQ, the environment is made up of many small isolated instances that, in combination, affect IQ. Rather large single environmental effects may also exist, but will occur idiosyncratically. Jensen also suggested that many of these effects are mediated biologically on IQ. The traditional psychological approach for studying the environment assumes a systematic and predictable path of action. Nonshared environmental effects do not appear to behave in this fashion.

GENETIC INFLUENCE ON ENVIRONMENTAL MEASURES

A second finding of major importance has been that measures of the environment are genetically influenced. Plomin and Bergeman (1991) reported that almost all traditional measures of the family environment are, in part, representing the genotypes of the people involved. Furthermore, they suggest that the relationship between environmental measures and IQ is in part genetically mediated.

Similar to this finding is the perspective that Scarr (Scarr, 1992; Scarr & McCartney, 1983) has espoused for some time; namely, that people, except in extreme cases, make their own environments and that the environment is therefore necessarily genetically affected. In its most extreme form, this line of reasoning holds that regardless of the source of the environmental influence, the impact on the individual must be filtered through that individual's perceptions, which are necessarily reflective of their genotype. In other words, the environment does not determine the individual, but the individual determines the environment.

One aspect of genetic influence that environmental researchers seem to support is that of genotype–environment interaction. I agree that intuitively it seems as if these interactions must exist; however, genotype–environment interactions are nonexistent in human literature (Bouchard, 1993; Capron & Duyme, 1989; Plomin, DeFries, & McClearn, 1990). I do not share Wachs' optimism about the use of molecular genetic advances for furthering our understanding of genotype–environment interaction. While I would like to believe that direct identification of genotypes with respect to IQ will make genotype–environment interactions readily apparent, due to the highly polygenic nature of IQ, I think that this problem will not be easily addressed. Identification of genotype–environment interaction assumes that the relevant environments and genotypes can be neatly measured and categorized. If very little direct evidence has been found that environmental influences on IQ behave in this simplistic fashion, then our chances of detecting genotype–environment interaction are minimal.

ACHIEVEMENT VS. IQ

As I read the chapters in this volume, I noticed that many of the authors in their discussions of the literature often equate achievement with IQ. I find this practice troubling. Phenotypic correlations between scholastic achievement and IQ scores are usually only around .50; thus, leaving 75 percent of the variance between the two unshared. Furthermore, behavioral genetic work has demonstrated that the amount of overlap in genetic variance between achievement and IQ is very high (Thompson, Detterman, & Plomin, 1991; Wadsworth & DeFries, 1993), suggesting that when there are discrepancies between IQ and achievement, the differences are due to environmental sources of variation. This work has identified an area on which environmental effects may be particularly important; yet, to my knowledge, no research has been conducted addressing this issue. As long as researchers treat achievement and IQ as identical phenotypes, the true effect of the environment on IQ will be confounded. Assuming that IQ scores represent an inherent, biologically based capacity to process information, scholastic achievement then represents the interaction between IQ and the knowledge and skills provided in a formal learning environment. By definition, the environment must play a greater role in determining achievement level as opposed to IQ.

WHERE DO WE GO NOW?

Behavioral geneticists have been successful at pointing out important leads on how the environment does and does not affect IQ; but, in general, they have been less than rigorous at following up on these leads. Similarly, many environmental researchers pay lip service to the role of genetics on the measurement and study of the environment, yet few have implemented studies that allow them to control for genetic effects. In fact, as environmental researchers fail to account for an appreciable amount of the variance in intelligence, the theories become more complex in an attempt to predict how the environment impacts on intelligence. As the theories become more complex, they become less and less amenable to empirical validation.

A great deal of research in developmental psychology and in behavioral genetics has involved samples of middle-class families where environmental action is predicted to behave in a systematic and direct manner. As I have previously discussed, environmental factors with the greatest influence probably act in a nonsystematic fashion and can be expressed at a biological level. Furthermore, I would argue that for the most part, environments within these samples are too homogeneous to "override" inherited tendencies (Scarr, 1992). It is only when extreme groups are studied, such as impoverished families where opportunities are quite restricted, that effects will be found. I would also suggest that extreme groups may also provide the only evidence for genotype–environment interaction (Bouchard, 1993; McGue, 1989).

In summary, to answer the question that I posed initially, "Why has the identification of environmental influences on intelligence been so elusive?", I would respond that for typical samples of children that are studied by developmental psychologists, the environment as has been traditionally defined does not affect individual differences in IQ. Under typical circumstances, a large portion of predictable variation in IQ is determined by the child's genotype, both directly through additive and nonadditive genetic effects, and indirectly through the selection, perception, and measurement of the environment. The remaining variance can be explained by random epigenetic effects, and random unpredictable environmental influences. These statements do not imply that environmental effects cannot affect IQ, nor do they imply that within deprived samples of children the environment does not have a limiting effect.

REFERENCES

Bouchard, T. J. (1993). The genetic architecture of human intelligence. In P. A. Vernon (Ed.), *Biological approaches to the study of human intelligence* (pp. 33–93). Norwood, NJ: Ablex Publishing Corp.

Bouchard, T. J., Lykken, D. T., McGue, M., Segal, N. L., & Tellegen, A. (1990). Sources of human psychological differences: The Minnesota study of twins reared apart. *Science, 250*, 223–250.

Capron, C., & Duyme, M. (1989). Assessment of effects of socio-economic status on IQ in a full cross-fostering study. *Nature, 340*, 552–554.

Chipuer, H. M., Rovine, M. J., & Plomin, R. (1990). LISREL modeling: Genetic and environmental influences on IQ revisited. *Intelligence, 14*, 11–29.

Jensen, A. R. (in press). The puzzle of nongenetic variance. In R. J. Sternberg & E. L. Grigorenko (Eds.), *Intelligence, Heredity and Environment*. Cambridge: Cambridge University Press.

McGue, M. (1989). Nature–nurture and intelligence. *Nature, 340*, 507–508.

Molenaar, P. C. M., Boomsma, D. I., & Dolan, C. V. (1993). A third source of developmental differences. *Behavior Genetics, 23*, 519–524.

Pedersen, N. L., Plomin, R., Nesselroade, J. R., & McClearn, G. E. (1992). A quantitative genetic analysis of cognitive abilities during the second half of the life span. *Psychological Science, 3*, 356–353.

Plomin, R., & Bergeman, C. S. (1991). The nature of nurture: Genetic influences on "environmental" measures. *Behavioral and Brain Sciences, 14*, 373–386.

Plomin, R., DeFries, J. C., & McClearn, G. E. (1990). *Behavioral Genetics: A Primer, 2nd Edition*. New York: Freeman.

Scarr, S. (1992). Developmental theories for the 1990s: Development and individual differences. *Child Development, 63*, 1–19.

Scarr, S., & McCartney, K. (1983). How people make their own environments: A theory of genotype–environment effects. *Child Development, 54*, 424–435.

Thompson, L. A., Detterman, D. K., & Plomin, R. (1991). Associations between cognitive abilities and scholastic achievement: Genetic overlap but environmental differences. *Psychological Science*, 2, 158–165.

Wadsworth, S. J., & DeFries, J. C. (1993). Etiology of covariation among measure of IQ and academic achievement. *Behavior Genetics, 23*, 568.

Chapter 18

Advances in Genetic Analysis of IQ Await a Better Understanding of Environment

Douglas Wahlsten

Department of Psychology, University of Alberta, Canada

Experts in human behavioral and cognitive development, exemplified by the authors of these seven chapters, frequently introduce genetics into their discussions, and they almost invariably import the specific brand of heritability analysis espoused by leading figures in the Behavior Genetics Association. However, in the larger world of behavioral and neural genetics, there are many of us who do not find this approach useful for understanding development and who regard the heritability coefficient as misleading or virtually impossible to measure without substantial bias in human beings. It is ironic that many excellent examples of organism–environment interaction and bidirectional causation presented in these chapters bolster previous arguments against heritability analysis (e.g., Gottlieb, 1991; Lewontin, 1974; McGuire & Hirsch, 1977; Roubertoux & Capron, 1990; Wahlsten, 1994), and this makes me wonder why developmental psychologists (e.g., Bronfenbrenner, Ceci, & Lenzenweger, 1993) take their genetics from a source that dismisses their carefully considered conclusions. The goal of human behavior genetics is to analyze phenotypic variance into components attributable to distinct genetic and environmental causes. Perhaps psychologists can relate very easily to this approach because they were weaned onto generic analysis of variance (ANOVA) and linear models at an early stage of their careers.

There are several methodological and conceptual shortcomings of human behavior genetics that severely limit the usefulness of variance partitioning. Consider a family where the offspring, all singletons, live with their birth parents. Because they share both genes (G) and environment (E), it is impossible to separate the contributions of confounded G and E to phenotypic resemblances within the family, and any number of plausible models could account for the data. Consequently, a special situation must be found where effects of G and E are separated in reality so they can be separated statistically.

Adoption after birth is often utilized for this purpose. Of course, the offspring lives in intimate association with its genetic mother before, during, and, for some time after, birth. Because prenatal environment can influence postnatal development (e.g., Busnel, Granier-Deferre, & Lecanuet, 1992; Miller et al., 1993; Morgane et al., 1993; Robinson & Smotherman, 1992), *the adoption method cannot separate the contributions of G and E*. A large portion of the behavior genetics literature is afflicted with an antiquated perspective on prenatal life, and investigators who correctly view adoption as a method to assess effects of post-adoption environment, not genotype, are the exception (e.g., Duyme & Capron, 1992). Experimental methods with laboratory animals can achieve a relatively clean separation of heredity and environment (Carlier, Nosten-Bertrand, & Michard-Vanhée, 1992), but these have not yet been employed in human research.

The difference between intraclass correlations of monozygotic (MZ) and dizygotic (DZ) twins will be purely genetic in origin only if the environments of the two types of twins are equally dissimilar. Numerous studies have found that the environments of MZ twins tend to be *more similar* than those of DZ twins, which implies their greater phenotypic resemblance could to some extent be environmental. Hence, *the twin method cannot separate the contributions of G and E*. Several behavior geneticists have addressed this difficulty by comparing a wide variety of measures of specific features of environment to the degree of difference in twin resemblance, and in many cases the correlations have not been statistically significant. Two shortcomings of this kind of test come to mind. First, let us apply the same logic to the genotype. Suppose there are 100 genetic loci scattered widely across the chromosomes, each of which exerts an independent effect of .5 points on the IQ score. The chances of one genetic linkage study detecting and then another study replicating an association of any one locus with IQ would be very small indeed, but this would not prove the genes are irrelevant for IQ and all variance is environmental. The most we can conclude from negative results for linkage is that no single genetic locus has a *major impact* on the phenotype in question (see Ginns et al., 1992).

Second, a decisive test of the importance of twin similarity in environment for IQ requires that we must know which specific aspects of the environment are relevant for IQ and how they combine to effect development of mental ability in singletons. Lacking this knowledge, how can we be so sure the measure we use is relevant to the question asked? It is clear from these seven chapters that a child's environment is dynamic, multileveled, interactive, and contextual. Simple measures of static environment remote from the thinking child are rejected by these developmental psychologists. Without exception, these authors state that the specific environmental effects relevant to the intellect are not well understood. If this is the case, how can behavior geneticists be so self-assured when they argue that the greater environmental similarity of MZ twins is unimportant for heritability analysis of IQ? And why do many developmental psychologists take the heritability coefficient so seriously when a foremost quantitative geneticist advises that "most of the literature on heritability in species that cannot be experimentally manipulated, for example, in mating, should be ignored" (Kempthorne, 1990, p. 139)?

Quantitative behavior genetics can advance further only when developmental psychology has learned more about the workings of environment. Will the two disciplines progress more rapidly by collaboration, as advocated by Pianta, O'Connor, and Wachs (this volume)? Many geneticists could certainly benefit from the lessons of developmental psychology. For example, authors of the Minnesota Twin Study (Bouchard, Lykken, McGue, Segal, & Tellegen, 1990) claim that there is an early critical period for environmental effects on IQ that makes adult experience irrelevant for their project and allows them to compute heritability from MZ twins reared apart but then reunited an average of *10.5 years* prior to IQ testing. However, no such critical period has been documented whereby the level of intelligence established early in life is automatically retained despite years of intervening experience (Ramey & Blair, this volume; see also Clarke & Clarke, 1976), and numerous examples of fluctuations in adult IQ are known. Likewise, the claim of Scarr (1992) that environment has an impact on intelligence only in extreme environments has no basis in empirical research. On the contrary, a more defensible general rule is that the larger the difference in environment, the larger the effect. Developmentalists comprehend this very well and consequently work with extreme environments whenever possible in order to increase the statistical power of their tests. Numerous mutations are known to impair intelligence (Wahlström, 1990), yet not one mutation is proven to produce genius. The fascinating chapter by Wagner and Oliver (this volume) helps to explain this perplexing pattern. Behavior geneticists working with twins and adoption will also benefit from careful attention to the issues raised by Flynn (this volume), because age cohort effects can substantially alter correlations between relatives when age is not equated for all subjects in the study. To date, the penetration of behavior genetics into developmental psychology has been remarkably unidirectional. Many benefits may now accrue if the flow of information is reversed until a better integration is achieved.

Perhaps the greatest shortcoming of heritability analysis is the inherent presumption that genetic and environmental effects on phenotypic development occur separately, without influencing each other. That is, there must be no GxE interaction. When such interaction occurs, it no longer makes sense to attribute definite percentages of variance to each causal entity.

Heritability analysis is typically used in situations where the investigator has no idea of which or even how many genes may influence a behavior. A decisive test of gene–environment interaction, on the other hand, requires that replicated genotypes be observed in different environments. When the relevant genes are unknown, there is no general method for equating genotypes in humans and the hypothesis of additivity is unfalsifiable. MZ twins do not solve this conundrum because there is no way to discern which twin pairs have the same genotype.

The bulk of evidence about the control of gene action in laboratory species indicates that *genetic effects are contingent and context-dependent.* Strict additivity of G and E must now be viewed as exceptional or fortuitous. Consequently, there is no defensible reason to posit a *general conceptual framework* involving separation of genetic and environmental causes. A discipline is built on insecure foundations when it presumes the validity of assumptions that related fields have already tested thoroughly and rejected.

Pianta and O'Connor (this volume) state that shared environment of two individuals can have different impacts for different genotypes. This undoubtedly correct view leads to difficulty with the concept of shared versus unshared E in twins. For a pair of MZ twins having the same G, exposure to a peculiar common experience will tend to affect them similarly and thereby augment the between-pair variance without changing the within-pair variance, and this will make estimated heritability appear to be higher. However, for DZ twins having different G, a common experience will tend to influence them differently when there is GxE interaction, and, depending on the kind of interaction, could inflate the within-pair variance disproportionally, which would also yield a larger measure of heritability. Thus, when G and E interact in certain ways, occurrence of a peculiar experience common to only some of the twin pairs tends to increase the apparent genetic component of variance and underestimate the impact of shared experience when the structural model assumes additivity.

Many behavioral and neural geneticists now eschew quantitative genetic analysis in situations where the genetic factors are poorly known and instead strive to comprehend the effects of specific genes interacting with each other and with environmental features of recognized importance. By analogy, the authors of this volume appear to be adopting a similarly fruitful strategy of analyzing specific elements of the global environment. They fret about the emerging complexity and the ability of current methodology to cope with developmental dynamics. Hopefully, the awesome complexity of ontogeny will not discourage them. Neurogenetics is still at the early stage of analyzing the brain-relevant genome and has only the faintest glimmer of understanding of how these genes work in unison as part of the nervous system. The realization that over 30,000 different genes, each expressed in a unique protein molecule, are expressed in the human brain (Sutcliffe, 1988), but that the functioning of only a few hundred is reasonably well known, has not stifled the growth of neuroscience. It is, however, a good prophylactic against the kind of hubris that has infected human behavior genetics.

REFERENCES

Bouchard, T. J., Lykken, D. T., McGue, M., Segal, N. L., & Tellegen, A. (1990). Sources of human psychological differences: the Minnesota study of twins reared apart. *Science, 250*, 223–228.

Bronfenbrenner, U., Ceci, S. J., & Lenzenweger, M. F. (in press). Nature–nurture reconceptualized in developmental perspective: Toward a new theoretical and operational model. *Psychological Review.*

Busnel, M. C., Granier-Deferre, C., & Lecanuet, P. (1992). Fetal audition. *Annals of the New York Academy of Sciences, 662*, 118–134.

Clarke, A. M., & Clarke, A. D. B. (Eds.). (1976). *Early experience: Myth and evidence.* New York: Free Press.

Carlier, M., Nosten-Bertrand, M., & Michard-Vanhée, C. (1992). Separating genetic effects from maternal environmental effects. In D. Goldowitz, D. Wahlsten, & R. E. Wimer (Eds.), *Techniques for the genetic analysis of brain and behavior* (pp. 111–126). Amsterdam: Elsevier.

Duyme, M., & Capron, C. (1992). Socioeconomic status and IQ: What is the meaning of the French adoption studies? *European Bulletin of Cognitive Psychology, 12*, 585–604.

Ginns, E. I., Egeland, J. A., Allen, C. R., Pauls, D. L., Falls, K., Keith, T. P., & Paul, S. M. (1992). Update on the search for DNA markers linked to manic–depressive illness in the Old Order Amish. *Journal of Psychiatric Research*, 26, 305–308.

Gottlieb, G. (1991). Experiential canalization of behavioral development: Theory. *Developmental Psychology, 27*, 4–13.

Kempthorne, O. (1990). How does one apply statistical analysis to our understanding of the development of human relationships. *Behavioral and Brain Sciences, 13*, 138–139.

Lewontin, R. (1974). The analysis of variance and the analysis of causes. *American Journal of Human Genetics, 26*, 400–411.

McGuire, T. R., & Hirsch, J. (1977). General intelligence (*g*) and heritability (H^2, h^2). In I. C. Uzgiris & F. Weizmann (Eds.), *The Structuring of Experience* (pp. 25–72). New York: Plenum.

Miller, R. K., Faber, W., Asai, M., D'Gregorio, R. P., Ng, W. W., Shah, Y., & Neth-Jessee, L. (1993). The role of the human placenta in embryonic nutrition. Impact of environmental and social factors. *Annals of the New York Academy of Sciences, 678*, 92–107.

Morgane, P. J., Austin-LaFrance, R., Bronzino, J., Tonkiss, J. Diaz-Cintra, S., Cintra, L., Kemper, T., & Galler, J. R. (1993). Prenatal malnutrition and development of the brain. *Neuroscience and Biobehavioral Reviews, 17*, 91–128.

Robinson, S. R., & Smotherman, W. P. (1992). The emergence of behavioral regulation during fetal development. *Annals of the New York Academy of Sciences, 662*, 53–83.

Roubertoux, P. L., & Capron, C. (1990). Are intelligence differences hereditarily transmitted? *European Bulletin of Cognitive Psychology, 10*, 555–594.

Scarr, S. (1992). Developmental theories for the 1990s: Development and individual differences. *Child Development, 63*, 1–19.

Sutcliffe, J. G. (1988). mRNA in the mammalian central nervous system. *Annual Review of Neuroscience, 11*, 157–198.

Wahlsten, D. (1994). The intelligence of heritability. *Canadian Psychology, 35*, 244–260.

Wahlström, J. (1990). Gene map of mental retardation. *Journal of Mental Deficiency Research, 34*, 11–27.

Chapter 19

Commentary: If the Nature–Nurture War is Over, Why Do We Continue to Battle?

Richard A. Weinberg

Institute of Child Development, University of Minnesota

With the naiveté of Pollyanna, I had come to believe that the war was over—so-called hereditarians and environmentalists, committed to undiluted polar positions about the sources of individual differences in intelligence, had laid down their swords, agreeing that the "either–or" philosophy about nature and nurture was untenable. But, indeed, there is strong evidence that the debate lingers on (e.g., Baumrind, 1993; Scarr, 1993; Wachs, 1993), resurfacing as predictably as the weeds in my garden.

Although social, political, and religious contexts have varied over history, and popular definitions and theories of intelligence have changed, the nature–nurture question remains about the same (Weinberg, 1989). Two essentially different approaches to understanding how individuals gain knowledge form the basis of this flourishing controversy: Locke versus Descartes, empiricism versus rationalism, a "blank slate" versus a "prepared mind," and behaviorism versus ethology (Spitz, 1986). At their extreme, hereditarian arguments defend notions of racial inferiority and supremacy regarding intellectual ability and question the credibility of social and educational intervention programs. At the far opposite end of the continuum, environmentalists have supported enrichment and intervention programs and social policies, guaranteeing the permanent "raising of intelligence" (Spitz, 1986).

The "either–or" philosophy has been fueled by misconceptions about the roles that genes and environments play *together* in affecting the development of intellectual skills, how people behave, what their measured IQs are, or how quickly they learn.

Confusion about the accounting of sources of variance in intellectual ability is perhaps due in part to a lack of appreciation of Allport's (1937) distinction between nomothetic and idiographic approaches to understanding human behavior. Psychology, including developmental behavior genetics, generally deals with broad, preferably universal, laws—it is a nomothetic discipline that focuses on laws of learning, perception, and cognition and is aimed at predicting the aggregate. Determining the sources of variance within a population including the heritability of characteristics is guided by a *nomothetic* approach. In contrast, an *idiographic* approach highlights the individual—and his or her idiosyncratic "system of patterned uniqueness" (Allport, 1937, p. 9). At the individual level, one cannot argue that 60 percent of a person's intelligence is due to their genetic background and 40 percent to environment. Such statements must be limited to descriptions of a population's variance. Furthermore, a heritability index cannot provide answers to questions about the etiology of an individual's skills or handicaps or the anticipated benefits of a particular intervention (Anastasi, 1971).

To understand the development of a particular individual—his or her ontogeny—requires an analysis of the unique circumstances—genetic *and* environmental history—of the particular individual. Indeed, the interplay of genes *and* environments is essential to produce development in any psychological domain (Rowe & Waldman, 1993). Furthermore, the individual's experiences play greater or lesser roles at different stages of development.

The recent identification of within family *nonshared environmental influences* (Plomin, 1990) has shed new light on our understanding of what makes *individual* children in the same family so different from one another. In turn, this has helped advance our understanding of environmental origins of individual differences in development. Similarly, in this volume, Wachs discusses Greenough's and Black's (1992) concept of *experience-dependent development*, encompassing neural processes involved in the storage of information *unique* to an individual.

As demonstrated in papers within this volume (e.g., Ramey & Blair), a wide range of environments can have a functional impact on development—change in the environment (e.g., early intervention) can shape changes in behavior—a phenomenon known as *malleability*. In fact, malleability (or plasticity) is a pervasive human quality throughout development (Lerner, 1984). However, malleability does not mean that given the same environment, all individuals will be affected the same way or behave alike. Individuals construct unique experiences from the same environmental opportunities as a result of their experiences and their genotypes (Scarr, 1993).

Furthermore, we should recognize that environmental effects can become cumulative organic (not genetic) effects. For example, although Down syndrome is a genetic condition that limits intellectual development, maternal drug ingestion and prenatal radiation are environmental effects that can produce organic damage, resulting in limitations to intellectual development (Horowitz, 1987).

Despite an expanded knowledge base and an increased repertoire of research methods, the rapprochement between the two "camps" of belief has been slow, in part because of a strong emotional climate laden with value and moral issues. We know from the social psychology of conflict resolution that perspective taking and interaction can facilitate communication between those who maintain disparate views. In that context, we (Waldman & Weinberg, 1991) have argued the necessity of incorporating specific environmental measures into developmental behavior genetic research and have called for fruitful collaboration between environmentally oriented researchers and behavior geneticists.

The current volume provides a rich agenda for such collaboration:

Flynn's discussion of IQ gains over historical time is a fascinating search for the environmental factors that must account for such changes. He argues that one must question, however, the assumption that IQ gains over time can be equated with intelligence gains. Flynn's macroview of environmental influences is related to the importance of "multilevel interrelated environmental domains" argued by Wachs. Focusing on "environmental systems," Wachs also documents the linkages between specific dimensions of the environment and domains of cognitive competence. However, he warns that even though environmentalists have gone beyond main effect theories to system-based approaches, these theories are descriptive rather than dynamic. They tend not to indicate how contextual and microenvironmental "processes are translated into individual variability in outcomes."

Pianta and O'Connor directly challenge the study of specific environmental effects by suggesting the concept of developmental niche and positing a role for the "concerted action of the organization of environmental influences." They propose sampling purposely from different niches and pursuing intervention studies within niches. In aggregate, such studies would allow us to observe how environmental genetic relationships vary across certain constraints.

Another perspective is provided by Borkowski and Dukewich, who suggest multiple ways in which attachment, through self-regulation, influences intellectual development. The direct and indirect effects of attachment on intellectual development are assumed to be magnified by macrolevel contextual factors.

Unfortunately, by focusing on *environmental* effects, the current volume reflects a continuing need to provide "equal time" to environmentalists in discourse about intellectual development. On the other hand, many of the papers do present a new research agenda and demonstrate an awareness of the barriers that have prevented conceptual détente. With a more vigorous commitment to sharing points of view and collaborating on research, perhaps an olive branch can be passed and a more balanced perspective achieved in describing the development of intelligence.

REFERENCES

Allport, G. W. (1937). *Pattern and growth in personality.* New York: Holt, Rinehart & Winston, Inc.

Anastasi, A. (1971). More on heritability: Addendum to the Hebb and Jensen interchange. *American Psychologist, 26,* 1036–1037.

Baumrind, D. (1993). The average expectable environment is not good enough: A response to Scarr. *Child Development, 64*(5), 1299–1317.

Greenough, W., & Black, J. (1992). Induction of brain structure by experience: Substrates for cognitive development. In M. Gunnar & C. Nelson (Eds.), *Developmental behavior neuroscience* (pp. 123–153). Hillsdale, NJ: Erlbaum.

Horowitz, F. D. (1987). *Exploring developmental theories: Toward a structural/behavioral model of development.* Hillsdale, NJ: Erlbaum.

Lerner, R. M. (1984). *On the nature of human plasticity.* New York: Cambridge University Press.

Plomin, R. (Ed.). (1990). *Nature and nurture: An introduction to human behavioral genetics.* Pacific Grove, CA: Brooks/Cole.

Rowe, D. C., & Waldman, I. D. (1993). The question "How?" reconsidered. In R. Plomin & G. E. McClearn (Eds.), *Nature, nurture, and psychology* (pp. 355–373). Washington, DC: American Psychological Association.

Scarr, S. (1993). Biological and cultural diversity: The legacy of Darwin for development. *Child Development, 64*(5), 1333–1353.

Spitz, H. H. (1986). *The raising of intelligence: A selected history of attempts to raise retarded intelligence.* Hillsdale, NJ: Erlbaum.

Wachs, T. D. (1993). The nature–nurture gap: What we have here is a failure to collaborate. In R. Plomin & G. E. McClearn (Eds.), *Nature, nurture, and psychology* (pp. 375–391). Washington, DC: American Psychological Association.

Waldman, I., & Weinberg, R. A. (1991). The need for collaboration between behavior geneticists and environmentally-oriented investigators in developmental research. *Behavioral and Brain Sciences, 14*(3), 412–413.

Weinberg, R. A. (1989). Intelligence and IQ: Landmark issues and great debates. *American Psychologist, 44*(2), 98–104.

Part 3

Replies to Commentaries

Chapter 20

Scots, the Physiological Correlates of IQ, and the Milwaukee Project

James R. Flynn

Department of Political Studies, University of Otago, Dunedin, New Zealand

The commentaries on my chapter focus on two themes: the causes of massive IQ gains over time and how we might improve IQ tests as measures of intelligence; the effects of massive IQ gains, particularly obsolete norms, and whether I have made mistakes in arguing that these necessitate reassessment of environmental factors. This response is confined to the four scholars who have made critical points.

THE BRAND HYPOTHESIS

Brand finds unconvincing my case against a genetic explanation of the 15-point IQ advantage of American whites over American blacks. Because my paper makes no mention of this subject, I will refer readers to the relevant publications so they can judge for themselves (Flynn, 1980, 1987b, 1987d, 1989, 1990a, 1992b). Brand also defends his own hypothesis about the major cause of massive IQ gains: IQ tests that encourage guessing either because of time limits or their structure. Going back into the past, every generation has supposedly been more scrupulous than the succeeding one. Therefore, its members have exhausted themselves by making sure each answer was correct and wasted scoring chances by refusing to guess on harder items. The significance of his hypothesis is that if it were true, we would have a diagnosis of how some IQ tests go astray as measures of intelligence.

Brand makes no response to the study summarized that tested his hypothesis and falsified it. Once again, Flieller, Jautz, and Kop (1989, pp. 11–12) analyzed a Binet-type test and found that poorer performance on *items completed* accounted for virtually all of the last generation's massive score deficit. According to Brand, their scrupulosity should have meant fewer mistakes on such items. Brand continues to place heavy weight on his misestimate of Scottish IQ gains between 1961–1962 and 1983–1984 on 68 unaltered items, mainly verbal, from the WISC (Brand, Freshwater, & Dockrell, 1989). Flynn (1990b) gave an estimate of 10 points and proposed a thesis about how Brand could have made the mistake of putting the gain at only 1.5 points. Brand never spelled out how he arrived at his estimate and, therefore, that thesis was conjectural. Since that time, Brand has never, either in his reply (Brand, 1990) or here, done anything except focus on that thesis. That is, he has never made one critical point against my method of estimating Scottish IQ gains.

I used the obvious method of scoring the children against the WISC manual. What with only 56 verbal items unaltered out of 116, that involves a number of steps (Flynn, 1990b, pp. 45–47) that cannot be summarized here. However, just to settle the question of whether gains were small or massive, I used a simple and direct method: a demonstration that the mean raw score of 10-year-olds from 1983 to 1984 matched that of 11.66-year-olds from 1961 to 1962; and that the later 13-year-olds matched the earlier 15-year-olds (pp. 43–44). Brand has never mentioned or challenged this analysis. Anyone reading this who recalls the old mental age method of calculating IQs will know that it simply settles the question of whether gains were 10 points or more.

Because Brand et al. (1989) admit that the Wechsler verbal is a test of his hypothesis, the Scottish data refute him. It is unclear why those data are so important because massive verbal gains in the United States are well-evidenced and suggested in Japan, Austria, and West Germany by weaker evidence (Flynn, 1987c, Table 17). Because Brand's hypothesis is false, the notion that the contentiousness of Asian Americans handicaps them on IQ tests is not persuasive. Indeed, because Chinese and Japanese Americans match whites on nonverbal tests and lag on verbal tests, their performance would be better explained, in Brand's terms, by assuming that they are less scrupulous than whites.

WHAT PROSPECTS PHYSIOLOGICAL CORRELATES

Concerning Jensen's comments, I must clarify that John Raven's estimate of raw score gains on Raven's Progressive Matrices is *not* based on simply projecting gains from recent times back to the late 19th century. He took the actual scores of subjects aged 20 to 65, tested in 1942, and subjects aged 20 to 70, tested in 1992, and plotted them by birth date. The oldest 1942 subjects were born in *1877* and the youngest 1992 subjects were born in *1977*, which gave him data covering 100 years (Raven, Raven, & Court, 1993, Graph G2). I converted raw score gains into IQ gains and used age 20 rather than birth dates, getting an estimated gain of 55 IQ points from 1892 to 1992 (this looks 5 years off, but it is really all right). The method assumes that a 70-year-old tested in 1942 would get the same score if tested as a 20-year-old in 1892. The possibility that those subjects would have scored worse in old age is one reason I suggested a minimum estimate of 40 points gained. That is almost certainly too conservative in that, as John Raven noted, the notion of a large performance decline with age was a product of cross-sectional data uncorrected for IQ gains over time.

Therefore, IQ gains really do go back 100 years, not Jensen's 50 or 60 years, and probably amount to at least 50 IQ points. As for the causes of massive gains, I cannot here comment on Jenson's lengthy and intelligent list of candidates, many of which I have analyzed elsewhere (Flynn, 1984a, 1984b, 1987a, 1987c, 1987e, 1987f, 1988, 1990b, 1992a, 1994). I agree with him that if we knew more about the biological substrate of mental development, we could assess whether that had altered over time and then assess whether there had been true intelligence gains over time. How close the physiological correlates of IQ—such things as reaction times, movement times, average evoked potentials (the brain's electrical response to stimulus)—take us towards that kind of knowledge is another matter.

There are two things the physiological correlates of IQ cannot do. First, they cannot tell us whether IQ gains over time represent intelligence gains. That question must be settled by the evidence emphasized in my chapter, whether or not real-world behavior signals enhanced intelligence. If there were signs of this, and RTs, MTs, and AEPs were stable from one generation to another, that very stability would count against them. The same is true of intervention programs. Take the Milwaukee Project: Whether the experimental children learned to read better than the control children was the best test of whether their IQ advantage was an intelligence advantage. If physiological score trends conflicted with the reading evidence, we would call them deceptive, just as we do IQ trends that conflict with real-world behavior evidence.

Second, physiological scores cannot evade meeting the criterion of within-generation external validity. If they are to replace, or even anchor, IQ tests, we cannot trust them because of correlations with IQ. We would have to ascertain whether physiological scores rise with age at least to maturity, whether they correlate with teacher estimates of ability, SES, upward mobility of sibling versus cosibling, and so forth—which brings us back to real-world behavior, that indispensable final count of appeal. What would be a reasonable level of external validity? I would suggest that physiological scores should at least match the predictive validity of Raven's. According to Jensen, Raven's measures *g* and little else, so a short-fall could not be explained on the grounds that the new test is more factor pure (Flynn, 1992a, pp. 348–350).

If physiological scores meet these two criteria, tally with real-world behavior evidence between generations and possess robust within-generation external validity, they will have the virtues of present measures of *g* and lack the flaws. Jensen (1991) is optimistic, partially because he considers RTs and MTs as lacking any intellectual content and unaffected by cultural factors. I am not so sure. Jensen himself has found that RT and MT studies across ethnic groups such as Chinese Americans, Anglo Americans, British, Japanese, and Hong Kong Chinese, show puzzling inconsistencies that pose the possibility of cultural influences (Jensen & Whang, 1992, pp. 408–409). My own analysis of the Lynn, Chan, and Eysenck (1991) data revealed evidence of cognitive strategies influenced by cultural factors. British children were faster on those RT/MTs that correlated with IQ among Chinese, and vice versa. Are we to conclude that British children have better Chinese brains than Chinese do, and that Chinese children have better British brains than British do? A more plausible explanation is that both groups had a "target plan"; that is, they fleetingly think about where to put a finger. The Chinese distribute that planning equally before and after they move off the home button, the British mainly before they move off the home button. Therefore, British have slower RTs than Chinese but faster MTs; and British get higher correlations between RTs and IQ (more cognitive loading there) and lower correlations between MTs and IQ (less cognitive loading)!

Perhaps we can disentangle RT and MT and eliminate the trade-off between them. However, this would not eliminate cognitive strategies or cultural influences, merely make their footprints less visible. I would like to see what happens to RTs when black children compete within the context of a "game" called point guard, where the goal is to get the basketball to the open player (odd man out) the quickest, and where a buzzer tells who won on each play. No takers for this research design as yet. In any event, the future and the evidence it brings will tell whether Jensen's optimism or my skepticism was prescient.

MILWAUKEE PROJECT: USE AND ABUSE

Garber and Hodge have criticized my reference to the Milwaukee Project, my comments on Philip Vernon, and the methodology of my book on Chinese and Japanese Americans. I want this scholarly exchange to have a positive outcome based on our shared commitment to the pursuit of truth. However, I will say this: In over 40 years, I have never accused a scholar of bias (to the best of my memory); I am not accusing anyone now; I do not intend to do so in the future. First, such an accusation adds no substance to critical points. Scedon, given its nature, I would feel I could not make even one serious mistake. Who could live up to such a standard? Not I and, as we shall see, not Garber and Hodge.

The half paragraph devoted to the Milwaukee Project makes a substantive point to be discussed in the next section. However, its wording reflects the history of a task undertaken a decade ago; namely, warning people about how obsolete norms can mislead by inflating IQs. Flynn (1984b) gave a wide range of examples, how we can go astray in assessing the reliability of tests, their standardization samples, their administration, and the effects of intervention programs, notably the Milwaukee Project. In the early 1970s, I had colleagues coming to me saying: "Arthur Jensen is so wrong when he says you can't boost IQ. Have you seen what the Milwaukee Project has done? They take black children at risk for mental retardation and give them IQs of 120." Textbooks began to refer to its dramatic results (Mussen, Conger, & Kagen, 1974, p. 354). My 1984 article lowered the IQs, led to much more guarded comment, and terminated at least one myth about the Project. At the time, I was quite aware that Heber and Garber were not making extravagant claims: As Clarke (1973, p. 15) said, "Heber exhibits proper caution in interpreting these data." Therefore, my 1984 analysis makes no assertion that would indict them. That is also true of a more full and sympathetic account written somewhat earlier (Flynn, 1980, pp. 182–187). However, after 1984, brief references reveal a slippage of precision over time. They do not say Heber or Garber believed in inflated scores—they refer to "the usual reaction" or say that the program "appeared" to raise scores—but some (usually the most brief) are subject to such an interpretation.

My persistence on the theme of obsolete norms may seem obsessive. In 1984, I thought that most would take the message on board and that the literature would be reviewed to isolate studies in which obsolete norms acted as a confounding variable. Some thinkers—Jensen, Lynn, John Raven, Garber himself—reacted immediately. In some areas, for example, studies using cross-sectional data as a measure of the effects of aging on IQ, there was a quick appreciation that generational IQ gains dictated a total reappraisal.

Others were slow to react. In New Zealand, Reid and Gilmore (1988) found that prisoners applying for parole were thought to have higher than average IQs because they were scored against obsolete norms. Whenever I saw a reference to Vernon's estimate of the nonverbal IQ of Chinese Americans, an alarm bell went off. If a sample of whites have their scores inflated above the white norm, no one will conclude that whites in general have a higher mean IQ than whites in general, because that is logically absurd. However, what about samples of Chinese and Japanese Americans? No one reanalyzed classic adoption studies like Skodak and Skeels. A book on the subject of IQ gains over time argued that adoptive homes had less impact on IQ today than in the pat, when the trend was in fact a creation of IQ gains (Flynn, 1993, pp. 560–561; Storfer, 1990, p. 63). I attempted to show that the whole question of whether we had a defensible criterion for mental retardation would have to be addressed. Thanks to IQ gains over time, the percentage of children eligible to be classified as such had fluctuated dramatically; therefore, the notion that we had accumulated a body of evidence in favor of any particular criterion lacked plausibility (Flynn, 1985). Scholars like Spitz responded. To this day, however, I have not found a test manual that tells psychologists the plain truth: that its new test, with its new and more demanding norms, multiplies the number eligible to be classified as mentally retarded overnight, sometimes by as much as a factor of 5 (Wechsler, 1992, p. 211).

Frustration brought repeated calls for reappraisal and these often used the Milwaukee Project as a sort of rallying cry: "Remember how some of us went astray when the first results appeared." I make no apology for not giving a summary of its purpose or design each time, but when using it to make a point, I had the responsibility not to say anything plainly mistaken. As Garber and Hodge point out, the sentence saying Heber and Garber accepted the inflated IQs at face value was false. Time simply dimmed memory. In addition, the assertion that they gave black ghetto children the advantages of an upper class home is my gloss, not theirs. Garber's (1988, p. 310) book describes the intervention differently: as giving children whose mothers had IQs below 75 a better microenvironment, one provided naturally by mothers with IQs above 100, although contending with disadvantaged circumstances. I am not certain my assertion was mistaken: Jensen (1989, p. 245) quoted Herber as quipping that the Project made the childhood environments of John Stuart Mill and Sir Francis Galton seem deprived by comparison, and Jensen concluded that the Project provided every cognitive advantage psychologists could possibly imagine. The wording, however, *was* mine.

Fortunately, the format of this volume allowed Garber and Hodge to correct me, so no reader of it will be misled. I am in their debt for that. To make certain that no one is misled, I have added a footnote to my chapter rewording the two sentences at fault. I cannot see that I have contributed much to the myths surrounding the Project—after all, the offending sentences have not yet seen the light of day.

MILWAUKEE PROJECT: AN ANALYSIS

That said, I have no inclination to qualify the substantive point of the passage in question. It argues that the cognitive stimulation of the Milwaukee Project lacked potency compared to the characterological traits inculcated by the Chinese American home. Garber and Hodge speculate as to whether I did my homework, read the book on the Project (Garber, 1988) and their reply to Jensen's critique (Garber & Hodge, 1989). I read both rather carefully, but extended comment to justify an example did not seem warranted. It does now.

Garber and Hodge indict Jensen, as they do a wide range of commentators including myself, for saying the Project aimed at "raising IQ," rather than preventing an IQ decline with age. Their preferred wording expresses the concern of those who see a gradual decline taking children below the IQ criterion for mental retardation. However, it has no operational significance for evaluating the effectiveness of the intervention. Their language and Jensen's language translate into exactly the same question: Did the experimental children end up with higher IQs than they would have, had the intervention never occurred? Both accept that the answer lies in analysis of the IQ gap between the experimental and control children. Garber and Hodge prefer to take them at age 7, right after they left the Project, when the gap was 22 points on the WISC (translated into the WISC–R scoring convention of norming on all races, the gap reduces to 20.7 points). Jensen prefers to take them at age 14, the last age for which IQ data was available, when the WISC gap was 10 points (translates into 9.3 points). There are arguments on both sides: for age 14, that the gap closes the farther the experimental children get from cognitive stimulation that mimicked the content of IQ tests; for age 7, that the Project cannot be held responsible for the gap closing after it was terminated.

Garber and Hodge believe Jensen should have noted that the IQ gap closed between ages 7 and 14 because the controls gained, not because the experimental group declined. That is not strictly correct: Translated into current scores, the experimental children went from a mean of 96 to 92. At any rate, however, the point is not central. If the gap closed because the experimental mean fell, this would show that the Project merely gave the experimental group a temporary benefit to be lost later; if the gap closed because the control mean rose, as it did, this would show that the Project merely allowed the experimental group to anticipate progress they would have made later anyway. The lesson is that no matter what description we give of the Project, whatever words we use to describe its purpose, what was done, what happened, still the gap between the experimental and control groups is the sole measure of its effectiveness. That fact must not be blurred.

The debate over the magnitude of the IQ gap is secondary to the debate over its significance. Does the IQ gap between the experimental and control groups represent a mere score difference, or can it be equated with an intelligence difference with real-world consequences? If a mere score gap, the Project did nothing towards salvaging children at risk from mental retardation, because it did not upgrade them mentally. Their inflated IQ scores would save them from being *classified* as mentally retarded, with a WISC–R mean of, say, 96 few would be at risk, but that is a different thing. The same objective could be accomplished simply by falsifying their IQ scores. If the IQ gap was an intelligence difference, the most important real-world consequence we would anticipate would be that the experimental children learned to read better than the control children.

Garber and Hodge plus Jensen all accept that over their first 4 years in school, the two groups show no statistically significant reading difference as measured by the Metropolitan Achievement Test. Jensen (1989, p. 254) believes that this means we cannot confidently equate the IQ gap with a *g* or intelligence difference. Garber and Hodge (1989) acknowledged that this "appears incongruent" (p. 292) but emphasize that intelligence contributes only 42% of achievement variance, the rest being explained by nonintelligence factors like "motivation, emotional stability, energy, persistence, work habits, interests, and values." Their reply shows why I had no hesitation in siding with Jensen. I can place only one interpretation on it: The Project really did enhance intelligence and this made a positive contribution to reading; the Project undermined desirable nonintelligence traits and this made a negative contribution; the two canceled out leaving no significant reading difference. I doubt anyone would assess the effects of the Project in these terms and to do so would almost certainly be false. The notion that the control children escaped characterological damage inflicted on the experimental children lacks all plausibility given the nurturing atmosphere created.

We can soften the analysis, *if* we ignore statistical significance, always risky, and take the nonsignificant reading the advantage of the experimental children at face value. Using 96 and 75 as the experimental and control means IQs, and using Garber and Hodge's value of .65 as the correlation between IQ and reading, we would expect a reading advantage of .91 *SD* (21 ÷ 15 = 1.40 *SD* 1.40 × .65 = .91 *SD*). Using the average Standard Score difference over Grades 1 to 4, and using an unattenuated *SD* of 15, the actual reading advantage was .41 *SD* (6.125 ÷ 15 = .41 *SD*). To account for the missing .50 *SD*, given a correlation of .76 between character and reading, we would have to assume a .66 *SD* character deficit (.50 *SD* ÷ .76 = .66 *SD*). The alternative to assuming a character deficit is to grant that only 45% (.41 ÷ .91) of the 21 point IQ advantage at age 7 was an intelligence advantage. This gives 9.45 IQ points (21 × .45) as the true intelligence advantage. And this is an almost perfect match for the IQ difference between experimental and control children that persisted to age 14! It is really a maximum estimate of the intelligence advantage because the experimental children almost certainly enjoyed a character benefit, rather than deficit, from participating in the Project.

Other factors complicate assessment of the Milwaukee Project. The experimental children may have had advantageous school placement and treatment (Jensen, 1989, p. 253); whether there was a statistically significant maths difference is uncertain (p. 255); fewer experimental children required special assistance or were held back at Grade 3; behavioral problems were similar for both groups (Garber & Hodge, 1989, p. 283). The best verdict is the Scots verdict of not proven. Here I do want to confess a bias; that is, a passionate desire that the Project would confer great benefit. Anyone who reads Flynn, 1980, p. 198 will sense the depth of feeling. My research into IQ gains over time helped me face facts: It too suggests that large IQ differences cannot always be equated with intelligence differences.

This analysis shows why the limited returns from the Milwaukee Project make a good counterpoint to the Chinese success story. The Chinese American home does not boost IQ, or avoid an IQ decline, whichever you prefer, but their children achieve in terms of adult occupations as if the mean IQ of Chinese Americans were 21 points above their actual mean (Flynn, 1991a, p. 101). Better to upgrade achievement and leave IQ alone, than upgrade IQ with less effect on achievement.

VERNON AND OBSOLETE NORMS

Turning to Philip Vernon, I had enormous regard for him. My book on Asian Americans is dedicated to his memory. It refers to him as a giant, a Columbus, the repository of my greatest debt, and says that he had only one limitation—"He could not know what would only be known in the future: that (Chinese and Japanese Americans) had inflated IQ scores because of obsolete norms" (Flynn, 1991a, Acknowledgments). I stand by that and by what is said in Chapter 2 of this volume: (a) Vernon thought Chinese Americans had a higher IQ than their white contemporaries; (b) studies in which their IQs were inflated by obsolete norms led him to that conclusion; (c) once that conclusion is abandoned, the problem of why Chinese and Japanese Americans outachieve white Americans cannot be solved by the presumption of higher intelligence and takes on a whole new complexion. My major purpose was to refute Vernon's conclusion, show that Chinese and Japanese Americans had no IQ advantage, and launch my own explanation of their overachievement. I have never attempted to give anything like a full account of Vernon's sophisticated and complex views on the subject, which of course do not neglect noncognitive factors. If we cannot correct an authority on a key point and go on to argue a new thesis, without a full account of his or her views, we are all in trouble.

Let me dismiss a preliminary point. I am said to be guilty of "casual scholarship" for citing David Brand, a *Time* journalist, to the effect that Jensen believed Asian Americans do well because they are smarter. Brand (1987, pp. 44–45) did not merely refer to Jensen's views; he summarized a discussion with Jensen about his studies of Asian Americans and his conclusion that they have a higher average mean IQ than whites. Perhaps as the son and spouse of journalists, I am prejudiced in their favor, but I have found good journalists as reliable as good scholars. In our correspondence, Jensen has referred to the significance of high Asian IQs. Last year, his kind words about my book over the phone included no rebuke that he had been misrepresented. I felt free to carry the reference over to my chapter in this volume.

Back to Vernon (1982, pp. 23–33). In a section entitled "Surveys by A. R. Jensen," he referred to these as "two extensive and recent studies." He summarized Jensen's results from children in Chinatown schools as showing that from Grade 3 on, their verbal IQs are little below 100 (the table gives an average of 97.09) and their nonverbal IQs about 110 (rounded off from the table's 111.14). This study, as Vernon presents it in the text, lacks any reference to a white control group. The Chinese are scored against IQ norms for all races (not whites only) that suffer from 16.5 years of obsolescence (Flynn, 1991a, p. 50). The major factor is the obsolescence. When all adjustments are made so as to score them against contemporary whites, the Chinese children get 88.6 for verbal IQ, 100.85 for nonverbal. Vernon's more complete discussion of these results in his appendix refers to no white control group. Jensen sent me his unpublished data and he used white children from Bakersfield, CA as a comparison group. Vernon may not have considered them sufficiently comparable to serve as such (Flynn, 1991a, p. 50).

Vernon then gave results from the Jensen and Inouye study of three ethnic groups—whites, blacks, and what he called Orientals (both Chinese and Japanese)—from the public schools of Berkley, CA. These children were actually tested in 1968. Vernon's reference to it as a recent study may indicate he thought it was done alter (it was unpublished). The IQs for all groups are inflated by the all races factor and obsolescence, here a matter of 15 years. Vernon is unaware of this and takes the inflated scores at face value, which gives Orientals 113 for verbal IQ, 118 for nonverbal. However, here the white comparison group had scores that were even higher, so Vernon is cautious. He noted that the school district is a superior one and merely described Orientals as close to white standards on nonverbal IQ and 5 points below on verbal IQ.

Finally, Vernon presents his summary values for Chinese American children, based on "more recent" studies: 97 for verbal IQ, 110 for nonverbal and spatial tests. These values are, of course, simply taken from Jensen's Chinatown study with its inflated scores due to obsolescence. Why was he not warned off by the Berkeley study, where Orientals did not exceed the white comparison group? Perhaps it was the Japanese admixture, but we will never know. What I do know is that these inflated values permeated the whole literature on Chinese IQ, due to Vernon's enormous and deserved authority. It was worthwhile correcting them. My values, based on the total relevant literature, were 97 and 100 (Flynn, 1991a, p. 64). These apply only to Chinese Americans born from 1945 to 1950, not to any prior or subsequent generation.

To my mind, the above simply settles the question of whether Vernon was misled by obsolete norms. The odd thing is that Garber and Hodge (Chapter

11, this volume) concede the point. They refer to Vernon's study as "based heavily on previously unpublished data provided by Arthur Jensen." They note that a majority of the studies Vernon reviewed did not involve scores inflated by obsolescence, but compared different ethnic groups, often Chinese and White Americans, taking the same tests at approximately the same time. This has no relevance if, as was the case, the studies on which Vernon based his summary values did suffer from obsolescence.

FLYNN AND OBSOLETE NORMS

In my book on Chinese and Japanese IQ and achievement (Flynn, 1991a), I divided Vernon's studies into early ones, before 1935, that do not apply to the postwar generation I analyzed 12 later studies, all those which use mainstream IQ tests. Some of the latter have Chinese subjects, some Japanese, some both. One I set aside as being made up of university students who can hardly be considered representative of the larger populations. Of the 11 remaining, 6 involve scores inflated by obsolete norms (Flynn, 1991a, Tables 2.5, 3.1, 3.3, 3.6, 3.7, & 3.8). There are often white comparison groups as well, and because correcting for obsolete norms is always chancy, where these groups seem appropriate, I use them as a standardization sample. I put their mean at 100 and score Chinese or Japanese Americans against them. Although Vernon often made comparisons, he did not, of course, do this. In every case of obsolete norms, *both* Chinese and whites are adjusted for obsolescence, either by actually compensating for obsolete norms, or by setting whites at 100, which of course amounts to the same thing. There is *not one instance* in which Chinese are adjusted whereas whites are not.

This is where Garber and Hodge make their worst mistake. They say that "Unfortunately, in Flynn's (1984-sic: read 1991) analysis, only the IQs of Chinese Americans, not white Americans, were adjusted. The scores of white Americans would have been equally inflated because they took the same tests." This is simply false. Worse, however, is the message conveyed. Here is a scholar arguing that Chinese appear to score above whites because of obsolescence: He adjusts the scores of Chinese so as to lower them, but neglects to lower the scores of whites—which, if done, would have left Chinese above whites and confirmed a Chinese IQ advantage. What could be more fallacious? Such as elementary mistake would render the whole book worthless.

For those who cannot take time to read it, I will note that several scholars who command respect have read it with care and missed this mistake. Thomas Sowell (1994) said that my "comprehensive survey of IQ studies over a period of several decades turned up no instance where Chinese or Japanese Americans . . . scored higher than the national average of white Americans" (pp. 180–183). Richard Lynn (1993, pp. 237–242) in his painstaking review missed it. They miss it because it is not there. Lynn challenges my values but has to sweat to do so; if I had made such an elementary mistake, he would have dismissed me in one line. In passing, I should make clear I still hold to my values. In my opinion, Lynn relies too heavily on three studies, the Grade 1 results from the Coleman report (he sets aside the results of all other Grades), another Chinatown study with only 53 subjects (Flynn, 1991a, Table 3.7), and a study done on the generation that came after the postwar generation I analyzed.

BALANCING THE BOOKS

I owe Garber and Hodge for saving me from contributing to a myth about their work. Now the debt is paid: I have saved them from bearing sole responsibility for originating a myth about my work. I will not speculate about the origin or motives of their mistake. For me, it is enough that we are all children of Adam and lavish more care on our own research than interpreting that of others. I am pleased that they endorse my analysis of Skodak and Skeels showing that the mother–child IQ gap was inflated by obsolete norms and that they used my formulae in adjusting scores in the Milwaukee Project. I take some pride in the fact that they could not have found those formulae elsewhere and that their scores from various tests would have been noncomparable without them.

REFERENCES

Brand, C. R. (1990). A "gross" underestimate of a "massive" IQ rise? A rejoinder to Flynn. *The Irish Journal of Psychology, 11*, 52–56.

Brand, C. R., Freshwater, S., & Dockrell, W. B. (1989). Has there been a "massive" rise in IQ levels in the West? Evidence from Scottish children. *The Irish Journal of Psychology, 10*, 388–394.

Brand, D. (1987, August 31). The new whiz kids: Why Asian-Americans are doing so well, and what it costs them. *Time*, 42–51.

Clarke, A. D. (1973). The prevention of subcultural subnormality: Problems and prospects. *British Journal of Mental Subnormality, 19*, 7–20.

Flieller, A., Jautz, M., & Kop, J-L. (1989). Les réponses au test Mosaïque à quarante ans d'intervalle [Answers on the Mosaique test after an interval of forty years]. *Enfance, 42*, 7–22.

Flynn, J. R. (1980). *Race, IQ, and Jensen.* London: Routledge.

Flynn, J. R. (1984a). IQ gains and the Binet decrements. *Journal of Educational Measurement, 21*, 283–290.

Flynn, J. R. (1984b). The mean IQ of Americans: Massive gains 1932 to 1978. *Psychological Bulletin, 95*, 29–51.

Flynn, J. R. (1985). Wechsler IQ tests: Do we really have a criterion of mental retardation? *American Journal of Mental Deficiency, 90*, 236–244.

Flynn, J. R. (1987a). Causal factors in generational IQ gains. *Nature, 328*, 765.

Flynn, J. R. (1987b). Flynn replies to Nichols. In S. Modgil & C. Modgil (Eds.), *Arthur Jensen: Consensus and controversy* (pp. 234–235). Lewes, Sussex: Falmer.

Flynn, J. R. (1987c). Massive IQ gains in 14 nations: What IQ tests really measure. *Psychological Bulletin, 101*, 171–191.

Flynn, J. R. (1987d). Race and IQ: Jensen's case refuted. In S. Modgil & C. Modgil (Eds.), *Arthur Jensen: Consensus and controversy* (pp. 221–232). Lewes, Sussex: Falmer.

Flynn, J. R. (1987e). The ontology of intelligence. In John Forge (Ed.), *Measurement, realism, and objectivity* (pp. 1–40). Dordrecht, The Netherlands: Reidel.

Flynn, J. R. (1987f). The rise and fall of Japanese IQ. *Bulletin of the British Psychological Society, 40*, 459–464.

Flynn, J. R. (1988). Japanese intelligence simply fades away: A rejoinder to Lynn. *The Psychologist, 9*, 348–350.

Flynn, J. R. (1989). Ruston, evolution, and race: An essay on intelligence and virtue. *The Psychologist, 2*, 363–366.

Flynn, J. R. (1990a). Explanation, evaluation, and a rejoinder to Rushton. *The Psychologist, 3*, 199–200.

Flynn, J. R. (1990b). Massive IQ gains on the Scottish WISC: Evidence against Brand et al.'s hypothesis. *The Irish Journal of Psychology, 11*, 41–51.

Flynn, J. R. (1991a). *Asian Americans: Achievement beyond IQ*. Hillsdale, NJ: Erlbaum.

Flynn, J. R. (1991b). Reaction times show that both Chinese and British children are more intelligent than one another. *Perceptual and Motor Skills, 72*, 544–546.

Flynn, J. R. (1992a). Cultural distance and the limitations of IQ. In J. Lynch, C. Modgil, & S. Modgil (Eds.), *Education for cultural diversity: Convergence and divergence* (pp. 343–360). London: Falmer.

Flynn, J. R. (1992b). Flynn replies to Lynn. In J. Lynch, C. Modgil, & S. Modgil (Eds.), *Education for cultural diversity: convergence and divergence* (pp. 378–382). London: Falmer.

Flynn, J. R. (1993). Skodak and Skeels: the inflated mother–child IQ gap. *Intelligence, 17*, 557–561.

Flynn, J. R. (1994). IQ gains over time. In R. J. Sternberg (Ed.), *The encyclopedia of intelligence* (pp. 617–623). New York: Macmillan.

Garber, H. L. (1988). *The Milwaukee Project: Preventing mental retardation in children at risk*. Washington, DC: American Association on Mental Retardation.

Garber, H. L., & Hodge, J. D. (1989). Risk for deceleration in the rate of mental development. *Developmental Review, 9*, 259–300.

Jensen, A. R. (1989). Raising IQ without increasing *g*? A review of *The Milwaukee Project: Preventing mental retardation in children at risk. Developmental Review, 9*, 234–258.

Jensen, A. R. (1991). Speed of elementary cognitive processes: A chronometric anchor for psychometric tests of *g*. *Psychological Test Bulletin, 4*, 59–70.

Jensen, A. R., & Whang, P. A. (1993). Reaction times and intelligence: A comparison of Chinese-American and Anglo-American children. *Journal of Biosocial Science, 25*, 397–410.

Lynn, R. (1993). Special review: Oriental Americans—their IQ, educational attainment, and socioeconomic status. J. Flynn: *Achievement beyond IQ. Personality and Individual Differences, 15*, 237–242.

Lynn, R., Chan, J. W. C., & Eysenck, H. J. (1991). Reaction times and intelligence in Chinese and British children. *Perceptual and Motor Skills*, *72*, 443–452.

Mussen, P. H., Conger, J. J., & Kagan, J. (1974). *Child development and personality* (4th ed.). New York: Harper & Row.

Raven, J., Raven, J. C., & Court, J. H. (1993). *Manual for Raven's Progressive Matrices and Vocabulary Scales* (section 1). Oxford: Oxford Psychologists Press.

Reid, N., & Gilmore, A. (1988). *The Raven's Standard Progressive Matrices in New Zealand.* Paper presented at Australian Council for Educational Research (ACER) Seminar on Intelligence, Melbourne, Australia.

Storfer, M. D. (1990). *Intelligence and giftedness: The contributions of heredity and early environment.* San Francisco: Jossey-Bass.

Vernon, P. E. (1982). *The abilities and achievements of Orientals in North America.* New York: Academic Press.

Wechsler, D. (1992). *WISC–III: Wechsler Intelligence Scale for Children–Third edition—Manual* (Australian adaptation). San Antonio, TX: The Psychological Corporation.

Chapter 21

Human Developmental Diversity: A Learning Society Perspective*

Daniel P. Keating

Ontario Institute for Studies in Education, Canadian Institute for Advanced Research

How *much* variance the environment contributes to human intellectual diversity is so limited a question as to direct our attention away from issues of far greater theoretical and applied significance. Knowing how much tells us little about *how* they interact dynamically to yield the observed diversity, and how we can most effectively *enhance* population competence. The structure of the current volume focuses attention on a strictly linear question, when we should

---

*The author is Royal Bank Fellow and Director, Human Development Program, Canadian Institute for Advanced Research, to whom he expresses appreciation for their generous support. Comments or questions should be addressed to the author: Centre for Applied Cognitive Science, Ontario Institute for Studies in Education, 252 Bloor Street West, Toronto, Ontario M5S 1V6. Telephone/Voicemail: 416 923–6641 (x2482); FAX: 905 688–6922 or 416 926–4708. Internet: DKEATING@OISE.ON.CA

be moving toward understanding the developmental dynamics. These dynamics involve the intertwining of biological and experiential forces across historical time, both phylogenetic and ontogenetic.

A further constraint on our discussion about this issue results from funneling it through the narrow measurement window of IQ. If competence is broadly multidimensional, then looking only at this academically derived variance obscures critically important competencies, especially those in the social domain.

Along with many of the contributors in this volume, I propose a moratorium on the bipolar and mostly fruitless debate. Instead, we may better direct research attention to the following handful of questions, implied in my chapter in Part 1 and refined by a reading of the commentaries:

1. How do genes and environment interact over the course of developmental history to yield the multifaceted diversity of competence and coping observed in human populations? An important key to where we may focus our attention arises from the behavior–genetic finding that most of the environmental variance occurs within rather than between families. This suggests that social interaction variables, such as specific parent–child relationships and exchanges, are more likely to yield interesting findings than are broad social address variables like social class. Findings from animal models, such as the work by Suomi cited in Chapter 3, emphasize how important the developmental dynamics are. Rhesus macaques bred to be genetically vulnerable (due to hyperreactivity) who were cross-fostered to highly nurturant and experienced mothers were found to be indistinguishable from adult macaques not at genetic risk on many physiological dimensions, and were slightly more likely to achieve higher than average troop status.

2. Whether intelligence is general or specific is the wrong question; it is always both. Human competence appears to have fractal qualities: Domain-specific variance is observable even in the most restricted samples (Matthews & Keating, in press); however, fundamental patterns of how we view and act in the world, interact with others, and learn and adapt to new experiences—what we have called habits of mind—permeate our everyday cognitive activity across many different domains. The core question in this context becomes how competence emerges from this rich dynamic between accumulating expertise and general habits of mind. For example, we need to understand more about basic patterns of emotion and attention regulation, which appear to be formed quite early in development and have substantial lifelong influence.

3. Recognizing this complexity—developmental dynamics across time and the multidimensional, fractal nature of human competence—seem sufficiently daunting that we are tempted to retreat to linear models which we understand better. Several commenters noted this complication. However, new methodological insights arising from within the general linear model, from our greater appreciation of historical methodology, and from our rapidly growing facility

with dynamic systems analysis offer substantial opportunities to address this complexity. We can aspire to a coherent account of human diversity without recourse to a premature reductionism.

4. We need to introduce the notion of population competence as a necessary adjunct to the notion of individual diversity of competence. The focal question of this volume is why some people are smarter than others: Is it in their genes or in the environment? At a population level, however, it is clearly the case that both social and technological competencies have expanded exponentially through both biological and cultural evolution (Keating & Mustard, 1993). A critically important question in this contemporary period of profound change is how to support and enhance broader and deeper *population* competence—in breadth, depth, and diversity.

Different views about this societal issue are at the heart of the nature–nurture controversy in the political domain. From a naturist perspective on human diversity, selection is the obvious social mechanism for enhancing population competence. From a nurturist perspective, access and equity are similarly obvious. These positions have mirror-image Achilles' heels as well. Selection implies marginalization of some segment of the population, which evokes social instability and constitutes a major drag factor for skill-based economies. When access and equity are inadequate supports for the development of competence across the population—which they often are because individuals with differing developmental histories require alternate pathways to competence, pathways not always supported by social institutions such as schools—the popular conclusion is that there is a lack of capacity, not just capability. It is clear that neither position offers much of value to central societal issues.

5. This leads directly to a key question: For what purpose are we attempting to expand our knowledge of human diversity in competence? A learning society perspective (Keating, 1995, in press) posits that we need all sorts of human ingenuity in the contemporary world. To foster this, we need to work simultaneously on these fronts: understand fundamental developmental processes that optimize outcomes across increasingly diverse populations; adapt social institutions to support this diversity of pathways for the full diversity of competence we require; monitor how we are doing, because circumstances change; and organize effective learning networks to support continuous improvement, to diffuse best practices, and to encourage wide discourse on these topics at many social levels and in many social contexts—the family, the school, the workplace, and the community. Our broad conceptual framework highlights the overall quality of the social environment as a central factor from this societal, historical, and population perspective (Keating & Mustard, 1993).

In pursuing this question, we need to be aware of the extent to which our theories and questions are driven by sociohistorical forces. A simplified contrast between the industrial and the information age (from Keating, in press) focuses on some of these patterns (see Table 1).

Table 1. Characteristics of Education in the Industrial and Information Ages

	Industrial Age	*Information Age*
Pedagogy	Knowledge transmission	Knowledge building
Prime mode of learning	Individual	Collaborative
Educational goals	Conceptual grasp for the few; basic skills & algorithms for the many	Conceptual grasp & intentional knowledge building for all
Nature of diversity	Inherent, categorical	Transactional, historical
Dealing with diversity	Selection of elites, basics for broad population	Developmental model of life-long learning for broad population
Anticipated workplaces	Factory models, vertical bureaucracies	Collaborative learning organizations

Note in particular how central the role of selection and assortment is to an industrial economy. Taking advantage of talent by advancing it preferentially works well when only a small elite are required for a growing economy and society. By contrast, taking advantage of population competence in a knowledge-based economy may be central to progress. This would need to be reflected in changes in many social organizations and institutions.

Of course, a learning society is only one possible future. An alternate future is one that yields a small skill/cognitive elite and a marginalized mass population. The long-term prosperity, or even stability, of this composition of population competence is hard to predict, but there are good reasons to be dubious about its prospects. In particular, the strong association between steep social gradients and negative outcomes in population health is a germane analogy (*Daedalus*, Fall 1994).

To achieve a broadly inclusive learning society requires that we attend to all developmental influences, and particularly to how they interact dynamically across time. We need to have a broad perspective on competence, and to adapt social institutions and practices to recognize the diversity of competence and the diversity of pathways for achieving it.

REFERENCES

Daedalus. (1994, Fall). Health and wealth. *Journal of the American Academy of Arts and Sciences* [Special Issue].

Keating, D. P. (1995). The learning society in the information age. In S. A. Rosell (Ed.), *Changing maps: Governing in a world of rapid change*. Ottawa, Ontario: Carleton University Press.

Keating, D. P. (in press). Educating for a learning society: The transformation of schooling. In J. Lupart, A. McKeough, & C. Yewchuk (Eds.), *Schools in transi-*

tion: Rethinking regular and special education. Scarborough, Ontario: Nelson Canada.

Keating, D. P., & Mustard, J. F. (1993). Social and economic factors in human development. In D. Ross (Ed.), *Family security in insecure times.* Ottawa, Ontario: National Forum on Family Security.

Matthews, D. J., & Keating, D. P. (in press). Domains and habits of mind: An investigation of developmental diversity in patterns of competence. *Journal of Adolescent Development.*

Chapter 22

*The Niche Revisited: Implications for Research on the Environment and Intelligence**

Robert C. Pianta

Thomas G. O'Connor

University of Virginia

In reflecting on the commentary and target chapters, we were struck by the continuing polarization of discourse in the study of intelligence. In a book devoted to research on *environmental influences* on intelligence, much of the commentary revolved around environment in relation to heredity. Efforts to tackle the serious methodological and conceptual issues regarding inquiry into environmental effects, raised in several of the target chapters, were largely

*This research was supported in part by NICHD grant R01HD26911 and NIDRR grant H133G20188. Address correspondence to the first author at: 147 Ruffner Hall, 405 Emmet St., Charlottesville, VA 22903.

ignored (Brand, Chapter 9, this volume) or oversimplified (Spitz, Chapter 16, this volume). Like Weinberg (Chapter 19, this volume), we also wonder why the nature *versus* nurture debate has become so reified, given the apparent widespread acceptance of gene–environment correlation and coactivity.

The evidence for gene–environment correlation is massive (e.g., Plomin & Bergeman, 1991; Scarr, 1992). Therefore, contributions to the study of intelligence must be evaluated, in part, on what they add tot his evidence. In this chapter we advance the concept of the developmental niche as a heuristic for better understanding environmental influences in the context of gene–environment correlation.

The concept of the developmental niche holds promise for understanding environmental effects on intelligence in two major areas of research: studies focused strictly on influences of environmental parameters, and studies examining environment(s) in relation to genetic influences. In the following sections we advance ideas proposed in our target chapter, reflecting on commentary pieces where relevant. As we stated earlier, we view the concept of niche as applicable to both the environment per se, and to the larger developmental system that encompasses both environmental and biological processes.

STUDIES OF ENVIRONMENTAL INFLUENCES

There are four areas of research related to environmental influences on IQ to which the construct of developmental niche can be useful (a) reconceptualizing the search for effects of specific environmental parameters, (b) the use of genetically informed samples in experimental studies of environmental impact (e.g., early intervention research) and individual differences research, (c) improving measures of the environment in genetically informative studies (e.g., adoption and twin studies) and (d) examining the development of niches (e.g., Flynn's work on secular trends, Chapter 2, this volume).

Niche and Specific Effects

We are not convinced that there is widespread, convincing evidence for the effects of a *specific* environmental parameter that can be measured and understood independent of other parameters (Spitz, Chapter 16, this volume). We stated our reasons for this in the earlier chapter and concluded to the extent there is a specific environment effect, it appears to involve contingency in interactions between the child and a wide array of proximal environmental agents (parents, siblings, learning materials).

For this reason, compiling a list of the environmental variables correlated with intelligence (or changes in intelligence) is not likely to advance understanding because of methodological and conceptual problems (notably, conceptual and operational overlap as well as problems in isolating specific effects from environmental systems). The Spitz list includes variables at many different levels, encompassing direct, indirect, and interactive effects, which, if entered in a regression, would account for relatively small percentages of IQ variance. This is one reason why we suggest that the common thread among these many variables, a thread that embraces their intercorrelation, is the niche that reflects their overriding organization. However, as we noted earlier, niches have many points of contact with the children they affect (proximal agents), these include interactions with caregivers, peers, aspects of the physical environment (toys, noise, lead).

We suggest that the niche functions as a "regulatory system" for the "parent–child" and "child" systems, and that at least two principles govern the extent to which the niche regulates development toward positive outcomes (in this case intelligence): (a) values of certain environmental features are kept within tolerance levels for which the organism is preadapted (e.g., noise and lead levels are not excessive, nutrition is adequate) and (b) contact with proximal agents that is contingent on the behavior of the organism allows for more optimal regulation. These principles embody gene–environment correlation within the niche, and nonspecificity of environmental effects. From these principles follow several implications.

There is much more to be learned about environmental effects on IQ from studies of particular "niches" in which the organization and arrangement of environmental forces from a number of subsystems are systematically examined. There is clear precedent for this type of analysis in the fields of anthropology, for example. Ethnographic analysis can be used to identify and isolate contrasting niches that can then be examined with the quantitatively oriented tools of developmental methodology, with respect to regulation of contingency of interactions between child and proximal agents. One particular strategy that could be utilized to a greater degree would be that of examining interaction effects of more distal and proximal environmental agents in order to reflect niche effects. Many environmental influences that have indirect effects on IQ (e.g., mothers' social support) operate within a niche to affect contingency of contact. In this context, it is possible to examine interactions of indirect or distal influence (maternal education, neighborhood hazards). These interaction terms can be entered hierarchically to remove niche influences or groups can be formed to reflect different niches and then the maternal sensitivity–IQ link examined within these combinations.

The search for specific environmental parameters is an approach to studying the environment that leads to arguing over very small pieces of the IQ pie based on R-square increments associated with various environmental constructs or measures entered in regressions. In any set of studies of environmental effects on IQ, a large number of "environmental factors" could account for small, but significant increments in R-square (Pianta & Egeland, 1994; Sameroff, Seifer, Barocas, Zax, & Greenspan, 1987). Scientists interested in the environment per se could argue over these small increments in a search for the one true environmental variable that accounts for the largest increment (Spitz, this volume). Within a developmental niche, specific environmental factors probably act a lot like individual genes. Each may have a small effect, additive or interactive. This idea of "poly-environmental" effects parallels the "poly-genetic" effects that are widely recognized within the study of genetic influences on behavior. Like for genes, it may be the interaction of lots of little environmental effects that each accounts for the "environmental variability" in IQ found across individuals. In this way, the niche captures the sum influence of these smaller effects.

Genetically Informed Samples in Studies of Environmental Impact

A range of studies and methodologies has been used to advance the notion that environments influence intelligence and cause changes in intelligence over time. One area of research that has been used in this regard is early intervention research (Garber & Hodge, this volume; Ramey & Blair, this volume). However, early intervention research was primarily designed to examine whether, and by what processes, certain outcomes of socio–cultural importance, could be affected by environmental manipulation (Garber & Hodge, this volume). The research agenda was primarily one of evaluating effects. Interestingly, early intervention research is used in arguments both for and against environmental (or genetic) effects. However, unless specifically designed to address gene effects, early intervention research cannot clarify this polarized discourse. In our target chapter we called for gene-informed early intervention research as a means to address this problem. Studies of early intervention programs on genetically informed samples could contribute to teasing apart limits of intervention influences, and reaction norms, for certain genetic and or environmental profiles.

With regard to typical individual differences studies of environmental influence on intelligence, the vast majority of these studies conceptualize and measure environments only through between-family designs. Individual difference studies, having nearly exhausted the range of environmental variables correlated with IQ, can now examine those (and other) factors in within-family designs. The niche-level perspective can be brought to bear on within-family analyses by examining the extent to which higher level niche influences (such as maternal social support or neighborhood factors) are differentially correlated with a particular within-family environmental factor (such as differential affection or treatment). That is, does a mother's experience of social support or the neighborhood in which the family lives affect her continent responsiveness to her two children, and is this difference reflected in sibling differences in intellectual performance?

Measuring the Environment in Genetically Informative Studies

The idea that shared environment corresponds only to shared familial environment fails to recognize that within-family differences (e.g., differential parental treatment) appear to account for the most of environmental effect on IQ in behavior genetic studies. Although embraced by behavior geneticists, this fact is often not reflected in behavior–genetic critiques of environmental research. For example, to suggest that "E" is irrelevant because of the high correlation between MZ twins reared apart (between family variance) does not indicate the extent to which within-family differences (such as in rearing environments) contributes to whatever differences in IQ exist between these sets of individuals.

On the other hand, adoption studies frequently are used to demonstrate that even within-family factors (differences between adoptive siblings) account for little variability in IQ. However, differences between adoptive siblings can be attributed to genes or within-family environmental effects. Measurement of the environment in adoption studies is often poor. Variables such as social address or living in the same home as an adoptive sibling do not adequately capture nonshared environmental effects that may relate to IQ. Again, to the extent that a more-encompassing perspective such as the developmental niche is recognized, then adoption studies will also recognize the importance of more comprehensive and detailed assessments of environmental differences between adoptive siblings.

Examining the Development of Niches

In a seminal paper, Sameroff (1983) forwarded the notion that environmental systems are themselves a product of evolution, and as these systems change, a corresponding dialectic process produces change in the organism. Likewise, the secular trends in IQ literature (Flynn, this volume) suggests that "development" of niches is correlated with corresponding and subsequent changes in IQ performance. Interestingly, as niches develop, heritability estimates change. This is most clearly seen in cross cultural studies of intelligence and achievement test performance (Stevenson & Lee, 1990) and in epidemiological-cohort studies; when opportunities for higher education are made available on a widespread basis, and integrated into the cultural norms (introducing considerably more environmental variability and new agents of influence), heritability estimates increase. Therefore, the level of environmental variability was related to the estimates of H, and as we argue in more detail in a later section, environmental variability was necessary to "produce" the apparent genetic influence.

In this sense even though H estimates may accurately reflect individual differences related to gene effects, these estimates vary according to the level of environmental variability available in the niche(s) studied. The more we know about development of niches, the more we can shed light on heritability.

STUDIES OF GENETIC INFLUENCE

We argued that the niche for the development of intelligence encompasses both biological and environmental processes. Having addressed implications of this perspective for studies of environmental influences, we turn now to examine implications for genetically or biologically informed research.

Niche Selection in Environmental and Genetic Research

The estimation of h^2 has been widely criticized as at best an academic, and at worst a helplessly erroneous endeavor. This sentiment reflects the lesson sometimes neglected in behavioral genetic research that the genetic and environmental estimates derived from analytic models are parameters. Heritability estimates vary across studies, and that this variability should itself be a focus of inquiry. Estimating the range of the genetic influence on intelligence is important because it helps establish bounds, and therefore is parallel in some ways to the hypothetical "reaction norm" discussed by Waddington and others. The task of subsequent research is to describe the variations in genetic influence associated with differences across niches and, in so doing, begin to discern which environmental influences appear to have the most promising effect in promoting intellectual development.

In the case of intelligence, unlike many other behaviors, there is remarkable convergence of genetic and environmental estimates derived from adoption, twin, and extended family designs using a wide variety of measures of "intelligence." By almost any standard, intelligence has passed a "robustness" test in genetic analyses. Nonetheless, variations in the genetic influence on intelligence—whether the estimate falls within a reasonably narrow range of .4 or .7—are worth considering. Rather than view the differences across methods and designs as entirely error (which is surely playing some role), we suggest that these variations reflect differences in the environment, or more specifically, the niche in which the individuals studied are embedded.

No one would doubt that without some environmental input, there would be no genetic influence on intelligence (at least because there would be little variability in observed intelligence). However, a dichotomy regarding environmental input seems to be an undercurrent in some of the debate exemplified in this volume: Either the environment is "good enough" or not "good enough." Further more, in the latter case (which unfortunately appears to be true for the many millions of persons living in poverty), genetic analyses may be misleading, presumably because the environment has been "insufficient" to "turn on" genetic forces (thus explaining the lower obtained genetic estimates found when these studies are conducted). Curiously, a similar interest in how variations in the environment affect genetic influence in the "normal" range of environmental experience is less of a concern even though, presumably, some environmental influences have sufficiently turned on genetic influences. Arguably, one might quite reasonably view research on specific environmental influences as an attempt to find the variations in the "good enough" environment that support, interact with, or otherwise overlay genetic influences. In either case, the obtained estimate for genetic influence contains information regarding the environmental resources, namely the developmental niche, and much of the debate regarding the genetic–environment influences on intelligence may benefit from examining how genetic influences on intellectual development vary across developmental niches.

Interestingly, there exists in the literature several examples of how the developmental niche qualifies the magnitude of the genetic parameter. In research on adoption, for example, Scarr and her colleagues (Scarr & Weinberg, 1983) found malleability—albeit in mild amounts—characterized the intellectual development of black children, or more to the point, disadvantaged children, adopted into advantaged (i.e., middle-class) homes. The consequent difference in heritability that would be obtained if these adopted children (into middle-class homes) were compared to their nonadopted siblings would reflect how different niches may promote, or restrict, genetic "potentials." A lower sibling similarity in intelligence would be observed than what would otherwise be expected from a purely genetic model.

In a parallel way, the genetic roots of achievement and the genetic overlap between intelligence and achievement is similarly expected to be defined or parameterized, by the developmental niche. For example, Thompson, Detterman, and Plomin (1991) reported that achievement and intelligence do share genetic influences and that environmental influences may be likely to explain nonoverlap. Therefore, just as genetic influences on intellectual abilities may change with the severity of the environment (e.g., Scarr, 1992), genetic influences on achievement—defined at the level of a change in school or at the level of accessibility of advanced education—will quite naturally influence the genetic parameter.

Differences in heritability across very disparate environmental conditions as noted earlier are not uncommon and not surprising, but require models of development that specify mechanisms of action. Heritability differences across less disparate environments such as levels of responsiveness studied by Bornstein and others are not available, but seem critical in order to refine our understanding of how specific niches qualify/influence genetic predispositions. Such studies would not only offer clues as to the kind of environmental input needed to foster intellectual growth, but also offer directions for gene–environmental interactions in which environmental influences may stimulate genetic change.

Gene–Environment Correlations and the Niche

As we argued before, the findings regarding gene–environment correlations are yet another example of how genetic analyses are analyses of a developmental niche. We discourage the unpackaging of "distinct" environmental influences because interesting environmental influences do not operate in a vacuum; that is, they are correlated. Gene–environment correlations may also work in less direct ways to confound main effects analyses. For example, consider the finding that temperamental factors such as task orientation, emotionality, and attention show genetic influence (see Plomin, Reiss, Hetherington, & Howe, 1994). Furthermore, parents respond differentially to sibling differences in genetically influenced characteristics; that is, there is a gene–environment correlation in child behavior/parental responsiveness (e.g., Lytton, 1977; O'Connor, Hetherington, Reiss, & Plomin, in press). This dynamic opens the possibility that parents would be more attentive to more task-oriented, attentive children and would therefore amplify the already significant genetic differences or predispositions in behavior and intelligence between siblings.

Of course, parents may also respond differentially to compensate for the less attentive and task-oriented sibling, but studies examining this alternative are unavailable. Naturally, in either case, the significant and meaningful gene–environment correlation in child behavior/parental responsiveness is an important qualifier to main effects analyses.

Evidence for Changes in Gene–Environment Influences

Much of the debate in this volume has taken place outside of a theoretical context that explains intellectual *development*. The most prominent commentary themes concerned issues of main effects versus interactions, definitions of intelligence, and cohort differences in intellectual performance. As we noted, many old issues and premises were redressed without regard to mechanisms that could illuminate discussions of available data. The absence of a coordinating theory and pertinent studies that incorporate genetic and environmental components will stall progress in this debate. For a volume on environmental influences on intellectual development, two processes seem especially important—environmental experiences provoking genetic change and environmental experiences that explain changes in intellectual behavior independent of genetic influences.

In regard to this first issue, environmental mechanisms for provoking genetic change seem illusory, but some interesting findings have been reported. For example, in an adoption study Cardon and colleagues (Cardon, Fulker, DeFries, & Plomin, 1992) found a significant genetic component to change in intellectual development when their subjects were 7 years old. Although specific environmental factors that may have altered the magnitude or nature of genetic influences were not examined in the study, the authors interpreted the novel genetic influence emerging at age 7 as perhaps a function of the introduction of school. Thus, Cardon et al.'s findings offer a hypothetical but perhaps specific example of how environmental experiences realize genetic influences. Cardon et al.'s suggestion that school may explain changes in the genetic influences on intelligence also opens up possibilities for examining how the genetic influence on intellectual development may change according to the nature and quality of the school or with the sensitivity or responsiveness of the home environment as studied by Bornstein and others.

Longitudinal behavioral genetic studies of intelligence offer an example of how stability or change in environmental influences may create change in intellectual development *independent of genetics*. For example, in the MacArthur Longitudinal Twin Study, Plomin and colleagues (Plomin, et al., 1993) found that change in intellectual development from 14 to 20 months—an age period that has received significant attention from environmental studies of intelligence—could be attributed to genetic and environmental influences. That is, some features of the environment shared by siblings (most likely a set of influences) created a (shared) change in their intellectual development *independent of genetic influences*. Nonshared environmental influences (which also include error of measurement) operated in a similar manner in promoting change and stability in intellectual development independent of genetic forces. It is interesting to note that shared environment is less salient in older children/adults (e.g., Cardon et al., 1992; Loehlin, Horn, & Willerman, 1990). A decrease in impact of shared environment over the first decade probably reflects the different intellectual trajectories followed by siblings and the decreasing similarity with which parents respond to them.

Finally, models of intellectual development are emerging in which environment and biology are integrated. Research in the neurosciences provides results on memory and learning, in which environmental input crates quantifiable differences in the brain (e.g., Greenough, Black & Wallace, 1987). The hypothesis that these processes are genetically mediated, such that some persons (at this point, animals, not people) may, for genetic reasons, be more responsive to environmental input requires further attention. Furthermore, there may be neurological differences between individuals showing higher levels of intelligence that may be explained via genetic routes, and individuals showing different intellectual abilities may be shown to be different genetically (Plomin et al., 1994).

Shared Environment vs. Nonshared Environment

The breakdown of environmental influences into "shared" and "nonshared" categories in behavioral genetic analyses has been an important and fruitful distinction. A shared/nonshared categorization of environmental influences spawned numerous research efforts which have since documented the wide variability in siblings' adjustment (Plomin & Daniels, 1987), the differential experiences of siblings within the same family—especially regarding parental differential treatment, and has invigorated attempts to incorporate behavioral genetics with family systems theories of personality development (Bussell & Reiss, 1993).

In addition to generating some intriguing analyses and hypotheses of family environmental influences on social, personality, and intellectual development, the shared/nonshared distinction appears to have created some confusion. Several of the commentaries mention the relatively small effects of shared environment in genetic analyses, and this finding is taken as evidence that environmental influences can at best have very small effects. To be sure, the hypothesis that genetic influences play a central role in the development of intelligence continues to receive overwhelming evidence across numerous research designs and various definitions of "intelligence." However, the parallel conclusion that environmental effects are relatively minimal because of the consistently small variance accounted for by the shared environment parameter does not follow.

As we noted in our target chapter, the shared environment parameter is apparently a proxy for only more "macro" types of environment (i.e., those influences that are shared by siblings *by virtue of living in the same house*, e.g., SES). Shared environment as defined by behavioral genetic models is not (necessarily) related to proximal processes such as maternal sensitivity that have been studied by researchers wishing to link individual differences in environmental input to individual differences in intellectual development. In fact, detailed observational studies indicate that parents do treat (e.g., respond to) their children differently, and that the magnitude of these differences corresponds to the genetic similarity of the siblings (e.g., Dunn & Plomin, 1986; Lytton, 1977; O'Connor et al., in press; Stocker, Dunn, & Plomin, 1989). Thus, a source of environmental influence may be familial, but it is probably more likely to be nonshared than shared by siblings. Studies of the degree to which parental sensitive and responsive scaffolding is "shared" by siblings requires further study. As we mentioned in our target chapter, available findings indicate that siblings may not share a developmental niche.

Similar confusion or ambiguity surrounds the discussion of the meaning of nonshared environment, which, as detailed in the commentaries, includes such diverse factors as prenatal influences and the *accumulation* of differential (extrauterine) experiences of siblings across many years. As is the case for the shared environment parameter, the nonshared parameter describes a type of environment effect rather than a specific effect. Notwithstanding the confusion that occasionally surrounds the meaning of nonshared/shared environmental influences, behavioral genetic studies offer compelling evidence that environmental influences impart a measurable impact on intellectual development.

In sum, we are optimistic that the emphasis on *versus* in the nature versus nurture debate can be reduced. We have argued that the concept of the developmental niche is a useful heuristic for generating studies that expand and hopefully clarify gene–environment correlation. Furthermore, this construct sets the stage for genetically informative studies of environmental influence (Bronfenbrenner & Ceci, 1993), as well as for reconceptualizations of environmental effects, that together can illuminate the complex processes related to intellectual development.

REFERENCES

Bronfenbrenner, U., & Ceci, S. J. (1993). Heredity, environment, and the question "how?"—A first approximation. In R. Plomin & G. E. McClearn (Eds.), *Nature, nurture, and psychology* (pp. 313–324). Washington, DC: APA Books.

Bussell, D. A., & Reiss, D. (1993). Genetic influences on family process: The emergence of a new framework for family research. In F. Walsh (Ed.), *Normal family processes* (2nd ed.). New York: Guilford.

Cardon, L. R., Fulker, D. W., DeFries, J. C., & Plomin, R. (1993). Continuity and change in general cognitive ability from 1 to 7 years of age. *Developmental Psychology*, *28*, 64–73.

Dunn, J., & Plomin, R. (1986). Determinants of maternal behavior towards three-year-old siblings. *British Journal of Developmental Psychology*, *4*, 127–137.

Greenough, W. T., Black, J. E., & Wallace, C. S. (1987). Experience and brain development. *Child Development*, *58*, 539–559.

Loehlin, J. C., Horn, J. M., & Willerman, L. (1990). Modeling IQ change: Evidence from the Texas Adoption Project. *Child Development*, *60*, 993–1004.

Lytton, H. (1977). Do parents create, or respond to, differences in twins? *Developmental Psychology*, *13*, 456–459.

O'Connor, T. G., Hetherington, E. M., Reiss, D., & Plomin, R. (in press). A twin-sibling study of observed parent–adolescent interactions. *Child Development*.

Pianta, R. C., & Egeland, B. (1994). Predictors of instability in children's mental test performance at 24, 48, and 96 months. *Intelligence*.

Plomin, R., & Bergeman, C. S. (1991). The nature of nurture: Genetic influences on "environmental" measures. *Behavioral and Brain Sciences*, *14*, 373–427.

Plomin, R., & Daniels, D. (1987). Why are children in the same family so different from one another? *Behavioral and Brain Sciences*, *10*, 1–16.

Plomin, R., Emde, R. N., Braungart, J. M., Campos, J., Corley, R., Fulker, D. W., Kagan, J., Resnick, J. S., Robinson, J., Zahn-Waxler, C., & DeFries, J. C. (1993). Genetic change and continuity from fourteen to twenty months: The MacArthur longitudinal twin study. *Child Development*, *64*, 1354–1376.

Plomin, R., Reiss, D., Hetherington, E. M., & Howe, G. (1994). Nature and nurture: Genetic contributions to measures of family environment. *Developmental Psychology*, *30*, 32–45.

Sameroff, A. J. (1983). Developmental systems: Context and evolution. In P. Mussen (Ed.), *Handbook of child psychology: Vol. 1, History, theory and methods*. (pp. 237–294). New York: Wiley.

Sameroff, A. J., Seifer, R., Barocas, R., Zax, M., & Greenspan, S. (1987). Intelligence quotient scores of 4-year-old children: Social environment risk factors. *Pediatrics*, *79*, 343–350.

Scarr, S. (1992). Developmental theories for the 1990s: Development and individual differences. *Child Development*, *63*, 1–19.

Scarr, S., & Weinberg, R. A. (1983). The Minnesota adoption studies: Genetic differences and malleability. *Child Development*, *54*, 260–267.

Stevenson, H. W., & Lee, S. Y. (1990). *Contexts of achievement. Monographs of the Society for Research in Child Development, Vol. 55*, (1–2, Serial no. 221).

Stocker, C., Dunn, J., & Plomin, R. (1989). Sibling relationships: Links with child temperament, maternal behavior, and family structure. *Child Development*, *60*, 715–727.

Thompson, L. A., Detterman, D. K., & Plomin, R. (1991). Associations between cognitive abilities and scholastic achievement: Genetic overlap but environmental differences. *Psychological Science*, *2*, 158–165.

Chapter 23

Nature versus Nurture, Again: Flogging a Dying Horse

Clancy Blair
Craig T. Ramey

University of Alabama at Birmingham

Judging from several of the responses to the first seven chapters of this volume, current topics in the study of intelligence have been current for quite some time. Accordingly, in our response we cover some old ground and some new ground while expanding on the parameters established in the Ramey–Blair developmental model of intelligence presented in Chapter 5.

In the model, both biological and environmental constraints are acknowledged and we focus on experiential history because, for us, this is where the action is. Certain respondents included in this volume have taken a developmental perspective (most noticeably in chapters by Douglas Wahlsten, Phillip Ackerman, and Lee Anne Thompson), and endorse the inclusion of experiential history as necessary for the accurate representation of the intelligence construct. This point, however, is one that is far from total agreement and those who insist on the primacy of nature in our understanding of intelligence are flogging a dying horse and suffer fro methodologic myopia. As always with nature versus nurture, the question is one of human development and we attempt here to reorient the current debate by emphasizing the developmental perspective.

The process of intellectual development is not well understood. This is made clear by a psychometric method that assesses intelligence by attempting to take development out of the picture by identifying a core of intellectual ability which is present from one developmental assessment to the next. We do not question the history of practical usefulness of the intelligence measures but feel that the psychometric tradition misses what is the most fascinating question of all, how developmental potential becomes actualized.

While evidence for *g* and evidence for the possible genetic basis for intelligence (separate but related areas of research) are interesting and informative, they are not only developmentally barren, but fundamentally at odds with current knowledge of human developmental processes. Further, as the role of experience in the conception of intelligence continues to give popular ground to theories stressing biological and genetic antecedents of development, there is the danger that an overly deterministic (i.e., simplistic) focus will result in models of development in which the role of genetics may be overemphasized (e.g., Scarr, 1992).

To begin, we believe intellectual development to be dynamic and characterized by change. A developmental perspective recognizes multiple influences operating over time and requires longitudinal research methodologies that can account both for individual level and group level determinants of change. Genetic and *g*-based research paradigms, on the other hand, seek the holy grail of behavioral science; namely, a single variable with large explanatory power. The genetic explanation is a maturational one that purports to "account" for substantial variation in IQ, but the question is only occasionally asked, variance in what? What does it mean to account for variance? Quite simply, when environments are homogenous, variance associated with heritability will be high and vice versa. This point merely serves to underscore the fact that statistical and methodological techniques are merely tools, not substantive explanatory constructs per se and for many, when given the behavior genetics hammer, everything looks like a nail.

Whenever large and overarching claims are made for a particular scientific methodology, it is time for a little perspective taking. Wahlsten's critique of the behavior genetics methodology is particularly strong both in suggesting the inadvisability of taking the heritability coefficient too seriously and in calling attention to the unidirectional flow of information from behavior genetics to developmental psychology. If nothing else, this chapter would like to contribute to a bidirectional flow of information from developmental psychology to behavior genetics, beginning with John Loehlin's four facts. It is indisputable that the correlations that provide the basis for the "facts" are part of the scientific record, but this does not render them immune from misinterpretation. The four facts are quite simply cornerstones in a current view of intelligence, a current paradigm to use the phrase with the intention of Thomas Kuhn. It is a paradigm that we are in opposition to and one which we will try to reorient with empirical evidence. In this case, however, reorient against a prevailing wind, a social and cultural climate which is in opposition to the view we advocate, and which, if Thomas Kuhn (1962) is correct bout the structure of scientific revolutions, makes our endeavor more than difficult.

The problem is that the grail is not at hand, intelligence is not written in the gene code, waiting there for us to puzzle out, and given current knowledge of human developmental processes and problems facing the estimation of *h*-squared, statements of the large percentage of variance in IQ accounted for by genes are without basis.

Along with Richard Weinberg in Chapter 19, we believe that the potential for behavior genetics and developmental psychology to speak past one another is contained in a miscommunication or misunderstanding concerning the distinction between individual differences in intelligence which may be stable and predictable and developmental functions which characterize groups of individuals and which may reflect experiential history. We have addressed this issue directly in two previous publications (Moser, Ramey, & Leonard, 1990; Ramey, Yeates, & Short, 1984) and will summarize only salient conclusions here. Some current problems in the study of the development of intelligence seem to ensue from misunderstanding about the meaning of *R*-squared and components of variability due to individual level characteristics and group level characteristics, or that due to genes and that due to environment. It is not that intellectual development is entirely driven by experience, just as it is clear that intelligence is not entirely the result of one's genes. These components of variation reflect differing but complimentary aspects of development that may make differing contributions to statistical models of development at different time points. This renders attempts for one side or the other to appeal to *R*-squared problematic.

That indices of the two components of variability in development are separable is clearly demonstrated in early intervention research which employs randomized designs. In an analysis of data from the Abecedarian Project, Ramey et al. (1984) demonstrated that individual differences in IQ are stable over time, and become increasingly stable and predictable with increasing age, while growth trajectories characterizing groups (i.e., educational daycare or not) reflect group experiential history. The important point is that the two components of variability are distinct and alteration of the group developmental function affects neither the predictability nor the stability of individual differences in IQ. Further evidence for the separate components of variance associated with experiential factors and child initial status factors in development has been provided by Blair, Ramey, and Hardin (1995) in an analysis of data from the Infant Health and Development Program (IHDP, 1990). Here, among children receiving the intervention, individual child and family background characteristics were found to predict IQ at single time points in development, but level of participation in early intervention, not background characteristics, was found to better account for developmental trajectory or change over time. To reiterate, the developmental function was not dependent on a child's relative standing with regard to initial status factors such as maternal education or birthweight, but was determined by experiential factors, by the amount of early intervention received. Further, no relationships between child and family background characteristics and participation in early intervention were detected.

The need to model individual and group-level variation in development is not a new issue. Traditional regression techniques using ordinary least squares estimation procedures, however, may not adequately model the two components of variability. More appropriate analysis of developmental data can be achieved through random coefficient or multilevel models utilizing maximum likelihood estimation techniques. Here, individual growth trajectories can be modeled and variation in the parameter estimate describing all individuals can be modeled at what can be considered a second or group level of analysis. This model-based approach to developmental data addresses two related analytical issues, contextual effects on outcome and problems associated with cross-level inference concerning individual and group effects on outcome. Essentially contextual effects are those which result from dependency in data, whether from repeated measures over time or from individuals nested within contexts. Early intervention research employing a randomized, controlled design provides a classic example of a contextual effect. That is, variation exists among individuals within treated and control groups as well as variation between groups. Problems with cross-level inference occur when group level characteristics are incorrectly used to make inferences about change at the individual level and vice-versa. When assessing the effect of these different levels on outcome, it is necessary to use an analytical technique that specifies appropriate error terms for within group and between group level tests of hypotheses (see Burstein, 1980, for a complete description of multilevel issues in data analysis).

Realizing that both individual variation and experiential history are important for development, we are left with the intriguing question concerning how they come together, how is behavior organized? How do enriched environments, (within our model) within a particular domain of knowledge, produce intellectual gains? The prevailing theoretical perspective for developmental psychology is a transactional one; transactional meaning that in the exchange between child and environment, change is produced in both. Data from an early intervention program for low birthweight, premature children known as the Infant Health and Development Program (IHDP, 1990) are suggestive of the transactional perspective. In the highly supportive environment of the educational daycare and family support provided by the intervention, largest treatment effects on tests of intelligence at 24 and 36 months of age were observed for children of mothers with less than a high school education. Conversely, children of college educated mothers in the control group scored as high as their treatment group counterparts at these same ages. These results are consistent with the theory that the transactional experiences outlined in Chapter 5 and which are known to characterize both the environment provided by early educational intervention and to exist more frequently in advantaged home environments, in part drive intellectual development.

Further, children in the IHDP intervention with birthweights < 2001 grams benefited on average 6.6 Stanford–Binet IQ points less at 36 months of age than did children with birthweights between 2001 and 2500 grams. An analysis conducted to determine whether this same effect for birthweight would be evident for control group participants in both more and less supportive environments, however, produced a slightly different pattern of results. For control group participants, birthweight was predictive of 36 month Stanford–Binet IQ only in the most supportive environment, consistent with the intervention finding, whereas no difference in IQ was associated with birthweight in the least supportive, highest risk environment, suggesting that something associated with LBW may be limiting children's abilities to benefit from developmentally stimulating transactions (Blair & Ramey, 1995).

A possible carrier mechanism for these effects may be social withdrawal. Lower LBW children in the intervention exhibited increased social withdrawal when compared with heavier children. this could represent a causal mechanism, aversion to stimulation that leads directly to slower progress through the developmental curriculum. If so, this would be consistent with the intervention literature demonstrating the effectiveness of early tactile and kinesthetic stimulation for low birthweight infants. Tactile stimulation was not a part of the comprehensive intervention offered by the IHDP or increased social withdrawal could represent a symptom of an underlying deficit, one associated with behavioral organization at the physiological level and limitations in intellectual development. Quite likely, however, organizational problems at the physiological level impact learning and attentional processes necessary for optimal development and result in reduced benefit from educational stimulation. Although the possible link between behavioral organization and intellectual development requires empirical support, one implication for research in this area concerns the co-occurrence of low birthweight and disadvantaged environment. In the highest risk environment, stimulation and support for learning are greatly reduced. As a result, the highest risk child within the most disadvantaged environment may not be identifiable until school entry and support for learning and environmental stimulation, to some extent, become available. With the low birthweight rate in the United States remaining at 7% accompanied by declining mortality, this delay in identification may result in problems in school for children who might have been helped through early intervention.

In conclusion, sustained attention to and consideration of the relationship between genes and environments has necessitated a reformulation of the behavior genetics paradigm in light of current theory and research in the study of human development. The challenge currently facing both early intervention and behavior genetics research is to link nature and nurture in research methodologies designed to assess the ways in which genetic potential becomes actualized. Only then can we begin to come to conclusions about the extent to which genotype and environment individually and synergistically contribute to the expression of intelligence both ontogenetically and phylogenetically. As Anastasi (1958) argued in her APA presidential address, the appropriate phrasing of the issue is how nature and nurture work together to produce individual differences at various points in development throughout the lifespan. Hopefully, keeping this in mind will put an end to protracted speculation concerning how much is intelligence the result of nature and how much nurture.

REFERENCES

Anastasi, A. (1958). Heredity, environment, and the question "how?" *Psychological Review*, *65*, 197–208.

Blair, C. & Ramey, C. T. (1995, March). *Do low birthweight and early intervention jointly determine developmental outcome?* Paper presented at the Gatlinburg Conference on Research and Theory in Mental Retardation and Developmental Disabilities, Gatlinberg, TN.

Blair, C., Ramey, C. T., & Hardin, M. J. (1995). Early intervention for low birthweight, premature infants. Participation and intellectual development. *American Journal on Mental Retardation*, *99*, 542–554.

Burstein, L. (1980). The analysis of multilevel data in educational research and evaluation. *Review of Research in Education*, *8*, 158–233.

Kuhn, T. S. (1962). *The structure of scientific revolutions*. Chicago: University of Chicago Press.

The Infant Health and Development Program (1990). Enhancing the outcomes of low birthweight premature infants: A multisite randomized trial. *Journal of the American Medical Association*, *263*, 3035–3042.

Moser, H. W., Ramey, C. T., & Leonard, C. O. (1990). Mental retardation. In A. E. H. Emery & D. L. Rimoin (Eds.), *The principles and practice of medical genetics* (Vol. II, pp. 495–511). New York: Churchill Livingstone.

Ramey, C. T., Yeates, K. O., & Short, E. J. (1984). The plasticity of intellectual development: Insights from preventive intervention. *Child Development*, *55*, 1913–1925.

Scarr, S. (1992). Developmental theories for the 1990s: Development and individual differences. *Child Development*, *63*, 1–19.

Chapter 24

Environmental Influences on Intelligence: Would That it Were That Simple

Theodore D. Wachs

Department of Psychological Sciences, Purdue University

The commentaries on the chapters in this volume neatly fall into three "fuzzy sets." First, there are several chapters where in the authors essentially focus on only one narrow issue, with only indirect relevance to the major issues that were raised by the original chapters (commentaries by Ackerman, Garber & Hodges, and Pollitt & Saco-Pollitt). These commentaries could easily stand by themselves, without reference to the original chapters. Without denying the potential importance of issues raised in these set of commentaries (e.g., the discussion by Pollitt & Saco-Pollitt on the role of parasitic infection in cognitive performance), the disjunction between the issues raised in these commentaries versus those raised in the original chapters does not encourage further integrative discussion.

The second set of commentaries essentially argue that the environment, as understood by the environmental researchers represented in this volume, is fundamentally unimportant and irrelevant for understanding individual variability in the nature or development of intelligence (Brand, Jensen, Loehlin, Lynn &

Thompson). My discussion of the points raised in this second set will focus on the multiple problems in the assumptions and database used by the authors of this group of commentaries.

The Brand commentary deserves special mention here, although not later. What makes Brand's commentary stand out is not the fact that it commits many of the general (*g*) conceptual and methodological errors described in the next section, as well as its own unique (*s*) set of errors; such as its seeming obliviousness to what is actually stated in the chapters (e.g., Brand argues that chapter authors ignore gene–environment covariance and gene–environment interaction, all of which were prominently featured in my chapter if nowhere else); its reliance on an absolutist view of heritability (see the commentaries by Thompson and Wahlsten for a clear rebuttle of this viewpoint); its condemnation of environmental researchers for ignoring relevant data, although Brand is clearly guilty in this regard (e.g., ignoring the anthropological work of Ogbu, 1990, on the role of cultural–environmental influences on black–white IQ differences), and its making wild claims based on single, problematic studies (e.g., the Lucus et al. study cited by Brand, which he claims shows that only breast feeding can boost IQ). Rather, what makes the Brand chapter stand out is its reliance on what I would call the Hyde Park style of debate; namely, the waker your argument, the more you deal with rebuttles or inconvenient facts by personal invective or insult. Rather than acknowledge the multiple weaknesses in his arguments (as documented earlier and later), Brand chooses to accuse environmental researchers of being "pusillanimous," using "gobbledygook," engaging in "massive repression" and "quack claims," "walling in . . . psychobabble" and being motivated only by a desire to obtain "research handouts." Given the tone of Brand's chapter it is not surprising that the sterile nature–nurture dichotomy continues unrelated. I will say nothing more about Brand except to hope that his future writings are directed exclusively to the Pioneer Fund.

The third group of commentaries acknowledges the meaninglessness of the nature–nurture dichotomy, and suggests way sin which we can enhance our understanding of the multiple processes that influence variability in cognitive development and performance. In this group are the commentaries by Clark and Clark, Spitz, Wahlsten, and Weinberg. The final portion of my commentary on commentaries both highlights and expand on the points the authors of these commentaries have raised.

DEJA VU: SOME MISLEADING CONCLUSIONS REVISITED

Environment and Biology

Jensen, Loehlin, and Lynn argue that real variability in intellectual performance is essentially driven by genetic–biological factors that influence the development of the central nervous system (CNS). For example, Jensen differentiates between "shadow" influences (presumably environmental), that impact only on test scores, versus "true" influences, that effect the biological substraits underlying intelligence. What these commentaries do not consider is the increasing body of evidence showing direct effects of the psychosocial environment on both biological "*software*" (e.g., brain biochemistry or patterns of hormonal regulation: Cacioppo & Berntson, 1992; Gunnar, 1994; Reite, 1987) and "*hardware*" (central nervous system growth and development: Diamond, 1988; Greenough & Black, 1992; Killackey, 1990; Reite, 1987). Along the same lines Gottlieb (1992) has presented evidence illustrating how environmentally induced changes in gene regulation can contribute to species changes in evolutionary pathways, well before there are any changes in gene frequencies.

The evidence presented earlier does not in any way deny the role of biological factors in intellectual development and performance. What it does illustrate is the role that environment may play in terms of influencing biological processes that, in turn, influence intellectual variability. Therefore, contrary to the conclusion drawn by Jensen, Loehlin, and Lynn, it seems clear that environmental influences can have a definite impact not only on test scores but also on "true" intelligence, via the impact of the environment upon biological development and functioning.

The Measurement of Environmental Influences

A number of the commentaries (Lynn, Loehlin, Thompson) base their conclusion of a lack of environmental influences on what has been described elsewhere as "social address" measurements of the environment (Bronfenbrenner & Crouter, 1983). Social address approaches use aggregate group classifications (e.g., social class, reared apart) as an index of an individual's environment. The inadequacies involved in using social address measurements as an index of individual environment have been previously and repeatedly documented (e.g., Bronfenbrenner & Crouter, 1983; Wachs, 1983, 1993). For example, both Lehlin and Thompson place heavy emphasis on the social address of being "reared apart." However, Bronfenbrenner (1986) has clearly documented how the degree of phenotypic similarity or dissimilarity between reared apart twins depends directly on the similarity or dissimilarity of the environments that these twins are encountering. Similarly, the social address term "reared together" is often taken to mean that children reared under the same roof are encountering essentially similar environments. The problems with this argument have been carefully delineated by Hoffman (1991), who demonstrated how children living under the same roof can easily encounter different objective environments, due to differences in the number or age of sibs, or different parental experiences with or different parental expectations for different sibs leading to differential treatment. In terms of measurement of the environment the bottom line now is the same as it was over a decade ago

(Wachs, 1983). To draw valid conclusions about the extent or nature of environmental influences one must directly measure the environment.

Statistics

A number of the commentaries (Loehlin, Lynn, Thompson) base their conclusion of very limited environmental influences on correlations between twins reared apart or unrelated adoptive children reared in the same family. I would refer the reader to an unfortunately overlooked paper by Weizmann (1971), who noted that the same study could provide evidence supporting or negating the role of environmental influences, *depending on the statistics that were used.* Given that the correlation is independent from the mean Weizmann clearly demonstrates how there is nothing inconsistent in a study showing significant correlations between genetically related children reared apart (genetic influence), while simultaneously showing significant shifts in the mean level of each child's IQ toward the IQ level of their foster parents (environmental influence).

By focusing on correlational evidence the commentators do indeed make a convincing case for the predominance of genetic factors over the environment. However, by focusing just on mean differences I can make an equally compelling case for the relative importance of family environment over genetic variables. By simultaneously estimating both means and correlations we come closer to reality, namely that while both environment and genes are necessary, in and of themselves they are *not sufficient* to account for the full range of intellectual variability (Wachs, in press).

Shared and Nonshared Environments

A number of commentators (Clark & Clark, Loehlin, Lynn, Thompson) argue that the only environmental influence worth considering is what has been called "nonshared environment." This conclusion relies heavily on two flaws discussed earlier; namely, the assumption that children reared under the same roof will be encountering the same environment (hence these children should resemble each other) and a reliance solely on correlations.

The commentary by Lynn further highlights an additional problem; namely, that all too often conclusions about shared and nonshared environmental influences are based on studies where the environmental contribution is derived from a *residual score*, with no attempt to actually measure the environment. The inadequacy of this strategy has been clearly documented in studies by Rose (Rose, Kaprio, Williams, Viken, & Obrinski, 1990; Rose, Koskenvuo, Kaprio, Sarna, & Langinvainio, 1988), which illustrates how conclusions about the extent of shared environmental influences depend directly upon whether one uses residual or direct measures of the environment. If residual scores are used then little evidence for shared environmental influences appears; if more direct measures of the environment are used then we find clear evidence for shared environmental influences. Again, this leads back to the conclusion that studies investigating the extent and nature of environmental influences must directly measure what they are investigating; namely, the environment (Wachs, 1983, 1993).

Organism–Environment Interaction

Both Loehlin and Thompson argue either that gene–environment interactions are seldom obtained in behavior genetic research or that these interactions account only for negligible individual variance in predicting IQ. Hence, they conclude that gene–environment interactions are a relatively unimportant source of influence on human intellectual development. This line of argument fails on both empirical and statistical grounds. Empirically, powerful interactions consistently appear in infrahuman and in human experimental studies (McClelland & Judd, 1993; Rutter & Pickles, 1991). Only in human nonexperimental studies is it difficult to find consistent organism–environment interactions. The methodological–statistical reasons underlying problems in detecting organism–environment interactions in human nonexperimental studies have been repeatedly discussed elsewhere. These reasons include (but are not limited to) the insensitivity of traditional statistical procedures for detecting interactions (Wahlsten, 1990), use of unstable or inappropriate measures of individual or environmental characteristics (Wachs & Plomin, 1991) and nonoptimal distribution of critical predictor or outcome variables in nonexperimental studies (McClelland & Judd, 1993). These formidable problems are often compounded when interactions operate only across a restricted range of populations or environments or when higher order interactions or organism–environment covariance is also operating (Wachs & Plomin, 1991). Methodological and statistical procedures for maximizing the probability of detecting existing interactions have been presented (e.g., McClelland & Judd, 1993; Wachs & Plomin, 1991), but rarely are these procedures utilized in the behavior genetic designs discussed by Loehlin and Thompson. Until more appropriate measurement and statistical procedures are utilized, conclusions about the extent and/or nature of organism–environment interactions must be considered *premature* at best.

Why are Things so Complicated?

In their commentaries Loehlin, Spitz, and Thompson all decry the complex nature of environmental influences and environmental theories, as postulated by environmental researchers. Why do these environmentalist make things so complicated is the refrain that regularly echoes from these commentaries. To the extent that I can speak for environmental researchers I can assure Loehlin, Spitz, and Thompson that we did not start out with the idea of creating complex models. Rather, the complexities that are described became apparent to environmental researchers only after extensively studying the nature of environmental influences. In the sense that Mendel was led by his data to the concept of recessive and dominant genes rather than to a more simplified blending model, so environmental researchers have been led by their data to an understanding that the complex structure of the environment requires complex environmental models. In this regard it is important to remember the criterion of parsimony does not require that we adopt the simplest model or explanation per se, but rather the simplest model that provides the best fit to the phenomena under study.

Two points raised by the commentators on this question deserve special consideration. First, Spitz presents a long list of potentially relevant environmental variables, and asks how we can progress by introducing more variables "to an already unmanageable list." This comment ignores the work of environmental theorists such as Bronfenbrenner (1993), who have provided us with a detailed description of the structure of the environment. Structural models of the environment, such as those of Bronfenbrenner, allow us not only to order a long list of environmental variables into specific domain categories, but also describes how these domain categories relate both to each other and to developmental outcomes (see Wachs, 1992). The end result is not an unmanageable list but rather a very manageable and detailed taxonomy.

In her commentary, Thompson argues that as environmental theories become more complex they become "less and less amenable" to "empirical validation." To those readers who might be tempted to accept Thompson's pessimistic conclusion I would recommend the recent work of Bronfenbrenner and Ceci (1993; in press). The environmental model developed by these authors encompasses many of the complexities that Thompson finds so bothersome, and yet still derives very specific and detailed predictions about when and how environmental influences should be manifest.

Nutrition and Intellectual Changes

Lynn's conclusion that improved nutrition is responsible for secular increases in intelligence is based on an outmoded model of nutritional influences which, among other things, ignores the clear covariance between nutritional and psychosocial environmental factors (Pollitt, 1988). His conclusion also ignores recent evidence indicating that the longterm effects of early nutritional supplementation primarily influence children from low sociodemographic groups, or that the cognitive improvements associated with early nutritional supplementation are accentuated by subsequent educational influences (Pollitt, Gorman, Eng, 1993).

Nonshared Influence are Mysterious and Unmeasurable

Thompson argues that nonshared environmental influences are unsystematic and unpredictable. As I have discussed elsewhere, the same environmental factors can either be shared or nonshared, depending on the extent of children's exposure to these factors (Wachs, 1992). As I have noted in my chapter, rather then being unsystematic or unpredictable, nonshared environmental influences are examples of the principle of *environmental specificity* The fact that specific environment development relations can be replicated both within and across cultures indicates that we are not dealing with unpredictable, random processes.

Measures of the Environment are Genetic in Nature

Thompson argues that measures of the environment are "genetically influenced." In her argument what Thompson is essentially discussing is nothing more than *organism–environment covariance.* Although it is traditional in behavioral genetic studies to assign organism–environment covariance processes to the genetic side of the equation, as has been repeatedly been pointed out (Oyama, 1989; Wachs, in press, Wahlsten's commentary, this volume), this approach is conceptually incorrect. Organism–environment covariance (and organism–environment interaction) is a unique source of variance that is *neither* genetic nor environmental, but rather is based on the combined contribution of two processes. Even if one concludes that covariance can be assigned to the genetic side because covariance might start with the genes (e.g., children with different genetically influenced intellectual characteristics may subsequently elicit differential stimulation from their caregivers), what must be recognized is that the origins of a developmental influence and the operation of the influence have no necessary connection with each other. For example, as noted by Rutter, Champion, Quinton, Maughan, and Pickles (in press), the reasons people choose to smoke cigarettes are quite distinct from the reasons underlying the association between cigarette smoking and lung cancer. Labeling organism–environment covariance as a "genetic" influence is misleading in terms of our understanding of the role played by genetic and environmental influences on intelligence.

CHARTING FUTURE DIRECTIONS

Both Lynn and Clarke and Clarke raise the question of whether the impact of family environments or early interventions remain stable across time. This question is part of the broader issue of what factors help to maintain, accentuate or attenuate the impact of prior experience on later development. Available evidence suggests that any of these three options can occur, depending on a variety of specific circumstance (Wachs, 1992). Potential circumstances that have been identified include the stability of salient environmental influences across time and settings (Bradley, Caldwell, & Rock, 1988), the degree to which exposure to prior experiences increases the probability of exposure to later experiences that act to maintain the influence of the prior experiences (bridging experiences—Clarke & Clarke commentary; causal chain models—Rutter et al., in press), or the degree to which later situations reflect or elicit earlier learned behavioral styles or vulnerabilities (Sroufe, Egeland, & Kreutzer, 1990). In addition, Bradley et al. (1988) have provided evidence indicating that the stability process that is operating may depend on the outcome measure under consideration.

At the risk of adding more complexity I would like to suggest the possibility of two other potential stability processes in addition to those suggested by Clarke and Clarke and others. The first involves the potential operation of active organism–environment covariance. The joint action of environmental and biological characteristics can produce individuals with specific phenotypic characteristics, who then seek out "environmental niches" that fit their individual characteristics. Stability of prior experiences can be maintained to the extent that individuals are successful in forming and maintaining appropriate niches for themselves. However, the formation and maintenance of appropriate niches is not solely a function of the individual per se. The current generation of environmental theorists (e.g., my chapter) have repeatedly stressed that individuals and their microenvironments (niches) exist in a complex structure of higher order environmental contexts, all of which have the potential to impinge on and influence niche formation and maintenance. Hence, it is critical for future research on environmental influences not only to look at the types of individual characteristics that are associated with types of preferred niches (e.g., Gunnar, 1994), but also to consider the degree to which higher order contextual factors, such as social stress, social support or cultural values, act to enhance or disrupt the individual's ability to form and maintain appropriate environmental niches for themselves.

The second potential alternative is specifically relevant to the question of the degree of "washout" of early cognitive intervention. Clarke and Clarke in their commentary rightly point to the potential influences on washout of non-intervention, higher order contextual factors. An alternative possibility, inherent in the commentaries of both Clarke and Clarke and Spitz, involves the influence of individual differences in reactivity to early cognitive intervention. For the most part, evaluation of cognitive intervention programs have focused on mean differences between treated and untreated children, ignoring the degree of *intraprogram variability in cognitive gains* following intervention. I would argue that the intervention programs that will be most susceptible to later wash out are those where there is high intraprogram variability in cognitive gains following intervention. This emphasizes the importance not only of looking at intraprogram variability, but also investigating the question of which individual characteristics moderate children's reactions to cognitive intervention efforts. One starting place to look for individual moderators is found in the discussion by Clarke and Clarke on "spontaneous advances."

Thompson raises a potentially important point in her argument that individual environmental factors, such as number of books in the home, may be of trivial importance. Thompson is correct to the extent that we confuse *manifest* with *latent* environmental variables. Isolated measures such as the number of books in the home, the number of times the parents focus the child's attention on a manipulative toy or the number of times the parents name colors the child is seeing on *Sesame Street* are manifest environmental variables. Taken in isolation, these variables probably will have little impact on a child's intellectual performance, if only because of the greater chance of measurement error associated with single assessments. However, individual environmental factors may be quite important for intellectual development when they are viewed as part of a class of latent environmental variables. For example, it is not the number of times a parent calls the child's attention to the novel features of an object within a short term observational session that influences intellectual development, but rather the repeated calling of the child's attention to different features of different objects across time. The operation of latent variable processes is not unique to the environment. In his commentary Wahlsten provides an excellent genetic analogy to the concept of latent environmental variables. Specifically, Wahlsten illustrates how the chances of detecting the influence on intellectual variability of a single genetic loci taken in isolation would be extremely small, although the combined action of multiple genetic loci might be quite powerful. Our ability to determine relevant environmental domains and our ability to develop effective environmental interventions will be maximized to the extent that our interventions or analyses are based on measuring or changing latent environmental variables rather than focusing on isolated manifest environmental variables. An excellent example illustrating both the measurement and the strengths and weaknesses of a latent variable approach to assessing the environment has been provided by Cook and Goldstein (1993).

In his commentary Wahlsten highlights the importance of a relatively understudied aspect of the environment; namely, the role of the prenatal environment. Wahlsten's discussion about the potential importance of prenatal environment is strengthened by infrahuman studies, suggesting that the operation of prenatal factors may provide a nongenetic means for intergenerational transmission of environmental influences (Dang, et al., 1992; Dell & Rose, 1993; Sackett, 1991). The possibility of cross-generational environment effects is also inherent in Weinberg's suggestion about the cumulative nature of environmental effects.

Finally, I am in complete agreement with both Wahlsten's and Weinberg's emphasis on the importance of developing collaborative research relations between genetic and environmental researchers. As noted by these commentators, both behavior genetic and environmental studies would benefit immeasurably by such collaboration. Potential domains for collaborative research have been noted both by Weinberg in his commentary, as well as by others (Bronfenbrenner & Ceci, in press; 1993, chapter in this volume). Certainly a good collaborative starting point would involve integrating direct and repeated measurement of proximal environmental microprocesses, as well as environmental macroprocess measures, into traditional behavior genetic twin and adoption studies. Although I firmly believe that such collaboration is essential I also share Weinberg's pessimism, in regard to both the lack of such collaboration in the past as well as to the factors that continue to block such collaboration (Wachs, 1993). From an environmental perspective real collaboration will be difficult until behavior genetic researchers are willing to use the conceptual and methodological approaches that environmental researchers have developed over the past 75 years. Unfortunately, it is precisely these conceptual and methodological advances that all too many commentary authors refuse to consider.

REFERENCES

Bradley, R., Caldwell, B., & Rock, S. (1988). Home environment and school performance. *Child Development*, *59*, 852–867.

Bronfenbrenner, U. (1986). Ecology of the family as a context for human development. *Development Psychology*, *22*, 723–742.

Bronfenbrenner, U. (1993). Ecological systems theory. In R. Wozniak & K. Fisher (Eds.), *Specific environments: Thinking in context*. Hillsdale, NJ: Erlbaum.

Bronfenbrenner, U., & Ceci, S. (1993). Heredity, environment and the question how—A first approximation. In R. Plomin & G. McClearn (Eds.), *Nature nurture and psychology* (pp. 313–324). Washington, DC: American Psychological Association.

Bronfenbrenner, U., & Ceci, S. (in press). Nature–nurture reconceptualized in developmental perspective. *Psychological Review*.

Bronfenbrenner, U., & Crouter, A. (1983). The evolution of environmental models in developmental research. In W. Kessen (Ed.), *Handbook of Child Psychology* (Vol. 1, 4th ed.). New York: Wiley.

Cacioppo, J., & Berntson, G. (1992). Social psychological contributions to the decade of the brain. *American Psychologist*, *47*, 1019–1028.

Cook, W., & Goldstein, M. (1993). Multiple perspectives on family relationships: A latent variables model. *Child Development*, *64*, 1377–1388.

Dang, D., Chaouat, G., Delmas, A., Hureau, J., Hidden, G., & Lassau, S. (1992). Assessment of factors responsible for variability of birthweight in the longtail macaque. *Foila-Primatologica*, *59*, 149–156.

Dell, C., & Rose, F. (1993). Differential maternal environmental experience prior to pregnancy: Effects across two generations. *Medical Science Research, 21*, 75–77.

Diamond, M. (1988). *Enriching heredity: The impact of the environment on the anatomy of the brain.* New York: The Free Press.

Gottleib, G. (1992). *Individual development and evolution.* New York: Oxford University Press.

Greenough, W., & Black, J. (1992). Induction of brain structure by experience. In M. Gunnar & C. Nelson (Eds.), *Developmental behavioral neuroscience* (pp. 153–232). Hillsdale, NJ: Erlbaum.

Gunnar, M. (1994). Psychoendocrine studies of temperament and stress in early childhood. In J. Bates & T. D. Wachs (Eds.), *Temperament: Individual differences at the interface of biology and behavior* (pp. 175–198). Washington, DC: American Psychological Association.

Hoffman, L. (1991). The influence of the family environment on personality. *Psychological Bulletin, 110*, 187–203.

Killackey, H. (1990). Neocortical expansion: An attempt toward relating phylogeny and ontogeny. *Journal of Cognitive Neural Science, 2*, 1–17.

McClelland, G., & Judd, C. (1993). Statistical difficulties of detecting interactions and moderator effects. *Psychological Bulletin, 114*, 376–390.

Ogbu, J. (1990). Cultural model, identity and literacy. In J. Stigler, R. Shweder, & G. Herdt (Eds.), *Cultural psychology* (pp. 520–541). New York: Cambridge University Press.

Oyama, S. (1989). Ontogeny and the central dogma. In M. Gunnar & E. Thelen, (Eds.), Systems and development (pp. 1–34). *The Minnesota Symposium on child Psychology, Vol. 22.* Hillsdale, NJ: Erlbaum.

Pollitt, E. (1988). A critical view of three decades of research on the effects of chronic energy malnutrition on behavioral development. In B. Schurch & N. Scrimshaw (Eds.), *Chronic energy deficiency.* Laussane: IDECG.

Pollitt, E., Gorman, K., Engle, P., Martorell, R., & Rivera, J. (1993). Early supplementary feeding and cognition: Effects over two decades. *Monographs of the Society for Research in Child Development, 58* (17), 1–99.

Reite, M. (1987). Some additional influences shaping the development of behavior. *Child Development, 58*, 596–600.

Rose, R., Kaprio, J., Williams, C., Viken, R., & Obrinski, K. (1990). Social contexts and sibling similarity. *Behavioral Genetics, 20*, 763–778.

Rose, R., Koskenvuo, M., Kaprio, J., Sarna, S., & Langinvainio, H. (1988). Shared genes, shared experiences and similarities of personality. *Journal of Personality and Social Psychology, 54*, 161–171.

Rutter, M., Champion, L., Quinton, D., Maughan, B., & Pickles, A. (in press). Origins of individual differences in environmental risk exposure. In P. Moen & G. Elder (Eds.), *Essays in honor of Urie Bronfenbrenner.* Washington, DC: American Psychological Association.

Rutter, M., & Pickles, A. (1991). Person–environment interaction: Concepts, mechanisms and implications for data analysis. In T. D. Wachs & R. Plomin (Eds.), *Conceptualization and measurement of organism–environment interaction* (pp. 105–141). Washington, DC: American Psychological Association.

Sackett, G. (1991). Toward a more temporal view of organism–environment interaction. In T. D. Wachs & R. Plomin (Eds.), *Conceptualization and measurement of organism–environment interaction* (pp. 11–28). Washington, DC: American Psychological Association.

Sroufe, A., Egeland, B., & Kreutzer, T. (1990). The fate of early experience following developmental change. *Child Development, 61* 1361–1373.

Wachs, T. D. (1983). The use and abuse of environment in behavior genetic research. *Child Development, 54*, 396–407.

Wachs, T. D. (1992). *The nature of nurture*. Newbury Park, CA: Sage.

Wachs, T. D. (1993). The nature–nurture gap: What we have is a failure to collaborate. In R. Plomin & G. McClearn (Eds.), *Nature, nurture and psychology* (pp. 375–394). Washington, DC: American Psychological Association.

Wachs, T. D. (in press). Genetic and family influences on individual development: Both necessary, neither sufficient. *Psychological Inquiry*.

Wachs, T. D., & Plomin, R. (1991). *Conceptualization and measurement of organism environment interaction*. Washington, DC: American Psychological Association.

Wahlsten, D. (1990). Insensitivity of the analysis of variance to heredity–environment interaction. *Behavior and Brain Sciences, 13*, 109–161.

Weizmann, F. (1971). Correlational statistics and the nature–nurture problems. *Science, 171*, 589.

Chapter 25

What Does the Nurture of Exceptional Performance Tell Us About the Nature of Intelligence?

William L. Oliver

University of Colorado, Boulder

Richard K. Wagner

Florida State University, Tallahassee

In our target chapter, we explored some of the implications of the research on exceptional performance for understanding environmental influences on intelligence. We speculated that principles underlying the development of extraordinary levels of proficiency also apply to the development of intelligence in children. Several of the comments on our article were skeptical of our claim that schooling plays an important role in the development of intelligence. Other comments were both provocative and constructive.

SCHOOLING AND INTELLIGENCE

We suggested in our target chapter that "the development of skilled performance, and of the broader repertoire of cognitive skills, knowledge and strategies that make up intelligence, represent the culmination of the actions, and interactions, of genetic and environmental influences over a prolonged period (Wagner & Oliver, this volume, p. 99)." Moreover, we suggested that "the kind of long-term practice provided by schooling appears to influence the development of subsequent intelligence. (p. 98). Loehlin (this volume) argues that these claims are not consistent with basic findings on intelligence.

Loehlin identifies four facts that he says should serve as "empirical beacons in the night" to guide our speculations about the environment and intelligence. Fact 1 (that identical twins reared apart correlate highly in their performance on IQ tests) in combination with Fact 2 (that genetically unrelated children reared together don't correlate appreciably in their performance on IQ tests) suggests that individual differences in IQ are almost completely due to genetic factors and nonshared variance. Fact 3 (that, for the general population, performance on the Vocabulary and Block Design subtests of the WAIS–R correlate about .52) attests to the general nature of intelligence. This fact could be buttressed by the massive body of research on general cognitive abilities. Finally, Fact 4 (that average IQ has been steadily rising in developed countries over the past 50 years) suggests that cognitive ability *is* influenced by the environment. An ancillary fact to Fact 4 is stressed by Loehlin: Average *verbal* IQ, presumably the most trained IQ component, has increased less than the other component measures of IQ. Loehlin argues forcefully that these facts pose problems for the view that intelligence as measured by aptitude tests reflects the practice and training that people experience in school.

When one tries to estimate the effects of heredity and environment on individual differences in aptitude, it is important to realize that one ends up with population parameters, not immutable physical constants. As Thompson (this volume) makes clear, environmental contributions will *appear* small if the environment is homogenous across the sample. In fact, as she further notes, research in behavioral genetics has mostly dealt with samples of middle-class families, which would be quite homogenous with respect to educational experience. If the samples of children used to derive the correlations for Facts 1 and 2 were heterogenous with respect to education, we would predict lower intraclass correlations for the identical twins reared apart and sizable correlations for the genetically unrelated children raised together. Because quality of education inevitably varies with socioeconomic status (SES), this pattern of results can partly be predicted on the basis of the 0.4 correlation of SES with IQ (Humphreys, 1992). Given these and other problems in interpreting correlational findings on individual differences in IQ (see also our target chapter and Wahlsten, this volume), it is especially important to consider other sources of evidence when one tries to assess the effects of schooling on IQ. Interestingly, Loehlin's Fact 4 provides such evidence.

The secular rise in IQ (Fact 4), as Loehlin acknowledges, could have been caused by improved schooling over the past 50 years. In fact, during this period more people have gained access to schooling, the mean number of years of schooling has increased, and, arguably, methods of education have improved. The secular rise in IQ becomes less surprising when related findings on schooling are considered. As we discussed in our target article, Ceci (1991) has identified a number of important effects of schooling on the development of intelligence. Given the converging evidence he assembles, it is tempting to view schooling as a necessary substrate for the development of intelligence. Of course, how quickly people learn and how much they benefit from education may depend on cognitive process that are influenced by their genes. Loehlin raises the interesting possibility that the extent to which one is willing to engage in deliberate practice in school settings may have genetic influences. On this view, similarity in IQ between genetically related people is partly due to their having similar inclinations toward school work. Interestingly, the amount of time one watches TV, which probably varies inversely with the amount of time one engages in school work, has been shown to have genetic influences in children under the age of five (Plomin, Corley, DeFries, & Fulker, 1990).

And what about the fact that verbal IQ has shown a lesser improvement than other "less-schooled aspects of IQ?" Loehlin remarked in his commentary that accounting for this fact would "require a bit of fancy footwork"; we will do our best with a little pedestrian reasoning. Greater exposure to schooling may have disproportionally improved nonverbal skills by encouraging analytical modes of thought and through direct practice of test-taking skills. At the same time, gains in verbal skills due to schooling may have been offset by changes in reading habits as people began to watch more television, both for entertainment and general information.

Finally, Loehlin's Fact 3 concerning the correlation between IQ subtests shows, as he argues, that tests of aptitude tap into generalizable skills, or component skills. That these skills are generalizable is further supported by the fact that composite measures of aptitude (especially *g*) predict both academic performance, and to a lesser degree, job performance (Hunter & Hunter, 1984). An important goal of early education is to identify and systematically train generalizable skills and knowledge so as to prepare people for various jobs or higher education. The training people undergo to develop exceptional proficiency in a given domain (e.g., chess, music, sports, etc.) is much more narrowly focused, and, consequently, the skills that develop do not transfer to other domains. Therefore, we believe that there are important similarities between the development of exceptional proficiency through training and practice and the development of intelligence through schooling, and that the differences one sees between the two types of learning has a lot do with the goals set by the educators and the learners themselves. Because generalizable skills are probably among the most difficult to develop, they cannot be acquired without a supportive environment and systematic practice over a prolonged period. As a review by Clarke and Clarke (1989) indicates, the only social and educational interventions that have succeeded have been those that exert their effects over a long period of time. Efforts to enhance learning during short, "critical" periods of life (e.g., Head Start) have been relatively unsuccessful.

IS THE RELATION BETWEEN ABILITIES AND PERFORMANCE STABLE?

Brand (this volume) cites evidence that "important human skills are loaded for *g* and not much else" and that there are correlations between cognitive measures and performance measures of skill learning tasks (Ackerman, 1987; Hunter & Hunter, 1984). We do not think that these findings are contrary to our main line of argument, as Brand asserts. As we discussed in our target article, a meta-analysis by Hulin, Henry, and Noon (1990) showed that cognitive measures of ability can do a fairly good job of predicting how well subjects will learn a new skill or perform on a new job. These measures, however, become increasingly poor predictors of performance over time. The meta-analysis by Henry et. al. included a wide range of longitudinal studies that examined the relation between abilities and performance over time. A meta-analysis by Hunter and Hunter (1984) also examined the relationship between ability and performance for various civilian and military jobs, but included only crossectional data, and, hence, gives no indication of whether the relationship between abilities and performance is stable.

SAVANTS

Clarke and Clarke (this volume) suggest that we consider the research on people with low IQs who show remarkable abilities in narrow domains. The tasks these so-called *idiots savants* or, simply, *savants*, learn early in life are often quite complex. For example, people with low IQs have been known to perform feats of calendar calculation (calculating the days of the week that arbitrary dates fall on), mental arithmetic (e.g., calculating cube roots of 6 digit numbers), and rapid memorization (e.g., memorizing a piece of music after a single hearing). Clarke and Clarke cite claims by O'Connor and Hermelin (1988) that the talents of savants often "emerge unbidden" without any systematic practice.

In support of this claim, Clarke and Clarke mention findings from a study of two 10-year-old calendrical calculators (O'Connor & Hermelin, 1992).

Other researchers who have reviewed the literature on savants have come to a very different view of the talent of savants (Howe, 1989; Treffert, 1989). These reviewers emphasize the importance of motivation and practice in the development of the savants' skills. Savants often have compulsive personalities, and often focus narrowly on the skills that they develop. Moreover, these skills (e.g., calendrical calculation) can usually be learned by college students after a month of practice (Ericsson & Faivre, 1988). THe major exception appears to be memorization of music, but Ericsson and Charness (1994) argue that this remarkable accomplishment reflects acquired skill. They point out that the savants who have exceptional memory for music are invariably blind musicians who must memorize music if they are to play an instrument. In addition, findings from studies of savants with exceptional memory for music (Charness, Clifton, & McDonald, 1988) show that they have difficulties memorizing unfamiliar, atonal music, suggesting that *acquired* knowledge about the structure of conventional music underlies their exceptional memory.

It is difficult to know how much weight to give to the findings of O'Connor and Hermelin (1992) that were very briefly discussed by Clarke and Clarke. This study examined the calendrical calculating skills of two 10-year-old savants, each with an IQ of 90, on four different occasions over a 4-month period. The savants' mean reaction times on a calculating task showed no speed up over the four sessions. No error rates, or statistics, were reported in the study, although the performance of one subject was characterized as error prone and "variable." Other findings from this study showed that the subjects used a sophisticated rule based on the fact that the Gregorian calendar repeats itself every 28 years. O'Connor and Hermelin concluded that their "results suggest neither practice nor talent alone can account for savant abilities and strongly emphasize the use of strategies in calendrical calculation" (p. 912). One must wonder whether the methods used in the study were sensitive enough to detect practice effects and, perhaps even more worrisome, whether the subjects were trying to improve through deliberate practice during the 4-month period. In order to show that calendrical skill was uninfluenced by practice, one would first need to monitor the amount and quality of practice in which the subjects engaged, and, second, need to assess possible improvement with experimental methods having reasonable statistical power.

As many have argued, the talents of savants provide a challenge for theories that stress the importance of general intelligence. How can people of low IQ (or *g*) learn such complex tasks, when, according to certain theories, they are presumed to have deficient information-processing abilities? After reviewing the literature on savants, some researcher shave taken an extremely negative view of the concept of general intelligence. For example, Howe (1989) concluded his book on savants with the following:

> Among all the findings, the point that emerges most forcibly, most repeatedly and perhaps most importantly is that many human abilities can operate with a greater degree of autonomy and independence than they are supposed to possess. The prevailing view that particular intellectual skills are constrained by an all-pervading general level of ability or intelligence has turned out to be something of a myth. (p. 168)

We note in passing that review of the literature on prodigies come to similar conclusions (Feldman, 1986). Prodigies are children who develop exceptional talent in such domains as sports, music, and science. The talents of prodigies do not appear to be strongly associated with high IQ or any other measure of aptitude (e.g., Jensen, 1990). Instead, the development of talent in prodigies appears to depend on an early commitment to a regimen of deliberate practice, usually under the supervision of an expert teacher (Ericsson, Krampe, & Tesch-Roemer, 1993).

A WOEFUL LAMENT: IF ONLY GPS COULD HAVE PLAYED BETTER CHESS

We suspect that Brand's (Chapter 9) (mean?) spirited critique reflects a frustration that goes well beyond the contents of the target chapters. It is reminiscent of the laments of others who perceived themselves to be facing an unfavorable paradigm shift.

In the late 1960s and early 1970s, Newell and Simon's General Problem Solver (GPS) was an ambitious computer-simulation model that simulated human problem solving. It was based on higher general (g_{comp} for computational g?) heuristics such as "means–ends" analysis that could be applied to problems from any domain. The view at the time was that when more powerful computers became available in the next decade or so, programs like GPS could compete with humans in intellectual challenges such as the game of chess. When it became apparent that the limitations of programs such as GPS were not a function of computational firepower but rather their limited knowledge base, the fields of cognitive science, cognitive psychology, and computer science took off in search of better understanding the acquisition, structure, and application of knowledge. Now knowledge-rich computer chess programs can beat all but the top players in the world.

A similar situation developed in the field of verbal learning. Researchers began to believe that most of what could be learned about learning from having undergraduates recall arbitrary paired associates was already known, and that the future—even with regard to progress about basic learning principles—was in *understanding how new knowledge interacted with existing knowledge* (see, e.g., Estes, 1982).

Meanwhile, as the fields of learning and cognitive science have taken off in new directions, a substantial part of the field of intelligence apparently remains stuck in the blocks, if not perhaps even moving backwards in a search for ever more information "contentless" tasks. Part of the problem appears to be an implicit belief among some that the field of intelligence proper should be limited to *the study of individual differences in IQ*. The upshot is that we know an awful lot about IQ correlates, but very little about much else, even including mechanisms responsible for the development of *g* from birth to adulthood. (After all, each of us fro our birth to adulthood demonstrates as much range in intellectual ability as is found between average adults and those afflicted with the severest forms of mental retardation.) Consequently, we are left with paradoxes such as the remarkable fact that the developmental peak of intelligence (i.e., fluid ability) occurs at about the time that actuarial tables indicate that young adults are most likely to kill themselves in automobiles, and well before we afford them privileges such as running for the office such as President of the United States or leading corporations.

Those who would limit the field of psychology to the study of individual differences in IQ risk relegating the field to the sidelines as the game of scientific discovery is played on the field, much in the way that areas of psychology that once dominated the mainland (e.g., verbal learning, behaviorism) became relegated to islands of more limited influence in the face of scientific advances.

CONCLUDING REMARKS

We are surprised by the strength of the conviction held by many people that the development of exceptional proficiency is largely determined by general cognitive ability. If general cognitive ability, as a fixed mental capacity, exerts its effects throughout he time a proficiency is developed, one would expect to see the following pattern of findings: (a) People with low aptitudes (IQ or *g*) could never learn tasks requiring complex information processings; (b) aptitude tests would be good predictors of ultimate performance of complex skills; (c) individuals (presumably with high aptitude) could become exceptional performers in such domains as music, chess, and writing in a few years instead of the usual 10 or so, and; (d) tasks used to assess aptitudes would be relatively uninfluenced by prior experience and could not be significantly improved by training. As we point out in our target article, however, a survey of the literature indicates that assertions *b* through *d* are false—aptitude tests became increasingly poor at predicting performance as skills are learned; exceptional performers require 10 years to achieve eminence; IQ is influenced by schooling, and tasks used to measure IQ (e.g., digit span) can be improved with specific practice. We briefly argued earlier, in response to comments by Clarke and Clarke (this volume), that assertion *a* is also false.

Science often benefits enormously by looking at extreme cases. Rare eclipses have provided astronomers and physicists with crucial data. Darwin's insights were helped along by his study of rare island species. Important theoretical advances in molecular biology resulted from the study of viruses, and there many other examples. Philosophers of science (Kuhn, 1970) have argued that major progress in science depends on accounting for rare or anomalous findings. Therefore, it is not surprising that early psychologists were drawn to the study of experts and eminent people as extreme cases. Both Galton and Binet put a lot of energy into the study of exceptional performance, which they sought to explain with their theories of intelligence. We believe that current researchers should also question whether their theories of intelligence can account for findings from the growing literature on exceptional performance.

REFERENCES

Ceci, S. J. (1991). How much does schooling influence general intelligence and its cognitive components? A reassessment of the evidence. *Developmental Psychology, 27*, 703–722.

Charness, N., Clifton, J., & McDonald, L. (1988). Case study of a musical monosavant. In L. K. Obler & D. Fein (Eds.), *The exceptional brain: Neuropsychology of talent and special abilities* (pp. 277–293). New York: Guilford.

Clarke, A. M., & Clarke, A. D. B. (1989). The later cognitive effects of early intervention. *Intelligence, 13*, 289–297.

Ericsson, K. A., & Charness, N. (1994). Expert performance: Its structure and Acquisition. *Psychological Bulletin, 49*, 725–747.

Ericsson, K. A., & Faivre, I. A. (1988). What's exceptional about exceptional abilities? In L. K. Obler & D. Fein (Eds.), *The exceptional brain: Neuropsychology of talent and special abilities* (pp. 436–473). New York: Guilford.

Ericsson, K. A., Krampe, R. T., & Tesch-Roemer. (1993). The role of deliberate practice in the acquisition of expert performance. *Psychological Review, 100*, 363–406.

Estes, W. K. (1982). Learning, memory, and intelligence. In R. J. Sternberg (Ed.), *Handbook of human intelligence* (pp. 170–224). New York: Cambridge University Press.

Feldman, D. H. (1986). *Nature's gambit: Child prodigies and the development of human potential.* New York: Basic books.

Howe, M. J. A. (1989). *Fragments of genius: The strange feats of idiots savants.* New York: Routledge.

Hulin, C. L., Henry, R. A., & Noon, S. L. (1990). Adding a dimension: Time as a factor in the generalizability of predictive relationships. *Psychological Bulletin, 107*, 328–340.

Humphreys, L. G. (1992). Commentary: What both critics and users of ability tests know. *Psychological Science, 3*, 271–274.

Hunter, J. E. & Hunter, R. F. (1984). Validity and utility of alternative predictors of job performance. *Psychological Bulletin, 96*(1), 72–98.

Jensen, A. R. (1990). Speed of information processing in a calculating prodigy. *Intelligence, 14*, 259–274.

Kuhn, T. S. (1970). *The structure of scientific revolutions* (2nd ed.). Chicago: University of Chicago Press.

O'Connor, N., & Hermelin, B. (1988). Low intelligence and special abilities. *Journal of Child Psychology and Psychiatry, 29*, 391–396.

O'Connor, N., & Hermelin, B. (1992). Do young calendrical calculators improve with age? *Journal of Child Psychology and Psychiatry, 33*, 907–912.

Plomin, R., Corley, R., DeFries, J. C., & Fulker, D. W. (1990). Individual differences in television viewing in early childhood: Nature as well as nurture. *Psychological Science, 1*, 371–377.

Treffert, D. A. (1989). *Extraordinary people: Understanding "idiot savants."* New York: Harper & Row.

Author Index

Subject Index